Learn
Science in 100 Words

Surendra Verma

About the Author

Surendra Verma is a journalist and author based in Melbourne, Australia. He has published numerous popular science books internationally, which have been translated into twelve languages. His recent books include:

in print editions

The Mystery of the Tunguska Fireball
Why Aren't They Here?: The Question of Life on Other Worlds
The Cause of Mosquitoes' Sorrow: Beginnings, Blunders and Breakthroughs in Science
The Little Book of Scientific Principles, Theories & Things
The Little Book of Maths Theorems, Theories & Things
The Little Book of Unscientific Propositions, Theories & Thing
The Little Book of the Mind: How We Think and Why We Think
Fools and Geniuses: a novel about science (and superstition)
and a children's book
Who Killed T. Rex?: Uncover the Mystery of the Vanished Dinosaurs

In Kindle editions

The Mystery of the Tunguska Fireball
Tunguska: A new look at an old mystery
Beginnings, Blunders and Breakthroughs in Science
Why Aren't They Here?: The Question of Life on Other
Worlds Fools and Geniuses: a novel about science (and superstition)
The Case of the Missing Electron: a science mystery for inquisitive children
The Little Book of Scientific Principles, Theories & Things
The Little Book of Maths Theorems, Theories & Things

In Apple Books edition

Who Killed T. Rex?: Uncover the Mystery of the Vanished Dinosaurs
The Mystery of the Tunguska Fireball

Contents

Introduction

Science in 100 Words presents short, simple and sparkling explanations of 100 important scientific concepts that are crucial to the understanding of science in the 21st century. Each concept is first described in exactly 100 words and then explained in detail in about 500 words.

In explaining science simply and succinctly, the book heeds Einstein's advice: 'Most of the fundamental ideas in science are essentially simple, and may, as a rule, be expressed in a language comprehensible to everyone.' At the same time, it is also guided by another of the great man's aphorism: 'Everything should be made as simple as possible, but not simpler.'

To make explanations simple and accessible to those without a background in science, the use of formulas and equations has been kept to a minimum. But explanations of some scientific concepts demand exact scientific notations and formulas. That should not turn off the reader. At first, they may seem intimidating, but they are as simple and comprehensible as the world's most famous equation, $E = mc^2$.

Happy reading!

1. Acid/Base

Acids and bases are two important groups of chemicals. Acids react with bases to form salts. Acids turn litmus (a vegetable dye) from blue to red, while bases turn litmus from red to blue. Most of us have tasted a lemon (which contains citric acid) and are familiar with the most common property of acids – their sharp and sour taste. Bases have a bitter taste. Acids are corrosive to metals, and bases have soapy feeling. Water solutions of both acids and bases are electrolytes; that is, they conduct electricity. A base which is soluble in water is called an alkali.

Remember that acids and bases are dangerous chemicals and they should, like all other laboratory chemicals, never be tasted or touched.

Chemists define acids and bases in many different ways. The simplest definition is that an acid is a substance which when added to water gives hydrogen ions, H^+. A base, on the other hand, gives hydroxide ions, OH^-. This definition is based on the theory proposed by Swedish chemist Svante Arrehenius in 1887.

$$HCl \quad = \quad H^+ + Cl^-$$
acid $\qquad$ acid solution

$$NaOH \quad = \quad Na^+ + OH^-$$
base $\qquad$ base solution

When water solutions of hydrogen chloride, HCl, and sodium hydroxide, NaOH are mixed we have

$$HCl + NaOH = NaCl + H_2O$$

The ionic equation of the reaction is

$$H^+ + Cl^- + Na^+ + OH^- = Na^+ + Cl^- + H_2O$$

As Na^+ and Cl^- are spectator ions, the reaction between an acid and a base can therefore be simply shown as

$$H^+ + OH^- = H_2O$$

This equation is called the ionic equation of neutralisation.

Though the Arrhenius theory provides a satisfactory explanation of many acid–base reactions, it's very narrow in scope. The modern atomic theory tells us that the hydrogen ion is a proton and therefore it cannot exist itself in water solution. It will react with the water molecule to form hydronium ions, H_3O^+, according to the equation

$$H^+ + H_2O = H_3O^+$$

The above ionic equation of neutralisation would now become

$$H_3O^+ + OH^- = 2H_2O$$

Interestingly, when Arrhenius discovered his theory of electrolytic dissociation, it was rejected by his colleagues. Even his professor of chemistry rejected it when it was submitted as part of his doctoral thesis. 'If sodium chloride is dissolved into sodium and chlorine, why does not a solution of sodium chloride show properties of the elements sodium and chloride?' he asked. The discovery of the electron in the 1980s proved once and for all that Arrhenius was right.

The Brønsted-Lowry theory, proposed in 1923, extends the definition to define an acid as a proton donor, and a base as a proton acceptor. This definition applies to systems which contain water as well as systems which do not contain water.

The strength of an acid or a base is a measure of the degree to which it shows acidic or basic behaviour. It is a property of the acid or the base and the solvent, and is independent of concentration.

Chemists use the pH scale to measure how acidic or basic a liquid is.

2. Antimatter

All elementary particles such as electrons, protons and neutrons have three characteristics: mass (some particles have no mass), positive or negative charge (some particles have no charge), and spin (every particle spins somewhat like a top). Each particle also has a mirror twin – antiparticle – with the same mass and spin but opposite charge. Antiparticles make up antimatter. When antimatter meets ordinary matter, they annihilate each other and disappear in a violent explosion in which mass is converted into energy as dictated by Einstein's famous equation $E = mc^2$, where E is energy, m mass and c the speed of light.

If you are a *Star Trek* fan you probably know that the *Starship Enterprise* is powered by antimatter. Antimatter is not the stuff of science fiction; it does exist.

As early as 1898, Arthur Schuster, a British physicist, suggested the fascinating idea that an exotic type of matter could exist with properties that mirror those of ordinary matter: 'If there is negative electricity, why not negative gold, as yellow as our own?' He added that this speculation was just 'a dream'. In 1928 the gifted British theoretical physicist Paul Dirac provided the mathematical basis for Schuster's dream. Dirac predicted that the electron, which is negatively charged, should have a positively charged counterpart: 'This would be a new kind of particle, unknown to experimental physics, having the same mass and opposite charge as the electron. We may call such a particle an anti-electron.' The symmetry between the positive charges in his theory also demanded an antiproton.

The discovery in 1932 of the anti-electron (now known as the 'positron', short for 'positively charged electron') in the cosmic radiation vindicated Dirac's bold prediction. Twenty-three years later, scientists at the University of California at Berkeley created the antiproton in a particle accelerator. We now know that every fundamental particle has an antiparticle.

If matter and antimatter annihilate each other, there is no likelihood of antimatter existing on Earth, or even in the solar system. The solar wind, the spray of charged particles emitted by the Sun in all directions, would annihilate antimatter. However, scientists speculate that antimatter could exist in other parts of the universe, but so far they have found no evidence. This has not stopped them from creating antimatter in the laboratory. In 1996, a team of scientists at CERN, the European particle physics lab in Geneva, created the first antihydrogen atoms. An antihydrogen atom would have a positron orbiting a single antiproton.

So now there is the experimental proof that antimatter does exist, what can it be used for? Because the annihilation of matter and antimatter creates enormous amounts of energy – in a collision of protons and antiprotons, the energy per particle is close to 200 times that available in a hydrogen bomb – it is tempting to look at antimatter as a potential source of energy. This energy might one day provide the fuel for interstellar voyages. The amount of antimatter required for space flights is unbelievably small. A few hundred micrograms could fuel a spacecraft to Jupiter, and the round trip would take only a year.

If you find all this a bit too far-fetched, then what about the idea of an anti-universe – a universe parallel to ours. Enter it and you will find your antimatter counterpart: antiyou. Don't shake hands – you'll annihilate each other.

3. Atmosphere

Earth's atmosphere is the gaseous envelope surrounding it. Up to about 75 kilometres from the surface the composition of the atmosphere is virtually uniform; however, it becomes thinner with increasing altitude. The boundary between the atmosphere and the outer space is generally regarded at about 100 kilometres. The dry atmosphere contains nitrogen (78.08% by volume), oxygen (20.95%), argon (0.93%), carbon dioxide (0.039%) and very low concentrations of neon, helium methane, krypton, hydrogen, nitrous oxide, carbon monoxide, xenon, ozone, nitrogen dioxide and iodine. However, the atmosphere is not completely dry: it contains 1 to 4% water vapour, mostly near the surface.

The temperature and pressure of atmosphere falls with increasing height. At an altitude of 10–12 kilometres at which jet aeroplanes fly the temperature is around –50°C.

The lowest layer of the atmosphere is called the *troposphere*, which begins at the surface and extends to a height of about 9 to 17 kilometres. All clouds and most of the dust and water vapour of the atmosphere are found in this layer.

The next layer, *stratosphere*, extends to about 50 kilometres above the Earth's surface. Its composition is like that of the troposphere but it's where ozone is formed and destroyed. This ozone layer shields the Earth from the Sun's harmful ultraviolet rays. The boundary between the troposphere and the stratosphere is called the *tropopause*.

Above the stratosphere is the *mesosphere*, which extends to about 80 kilometres. The boundary between the stratosphere and the mesosphere is called the *stratopause*.

The region above the mesosphere, in which temperature increases with height, is called the *thermosphere*. The boundary between the mesosphere and thermosphere is called the *mesopause*.

The outermost region of the atmosphere, from about 400 kilometres, is called the *exosphere*.

Ionosphere is the ionised layer of the atmosphere. It extends from about 50 to 1000 kilometres and contains ions and free electrons, which are caused by solar radiation. The ionosphere reflects certain radio frequencies and thus enables radio transmissions to be made round the curved surface of Earth. Sunspot activity can cause abnormal variation in the ion density in parts of the ionosphere and thus cause communications blackouts.

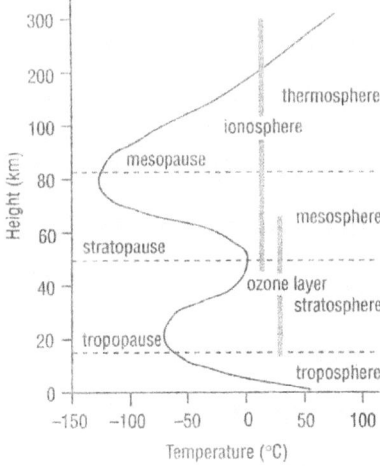

4. Atom

An atom is the smallest particle into which an element can be broken by chemical action and which can take part in a chemical reaction. Each atom consists of two parts: (a) a positively charged nucleus containing protons and neutrons; and (b) a cloud of negative charged electrons, surrounding the nucleus. The nucleus is very small and dense. Most of the mass of the atom is concentrated here, but the volume of the nucleus is extremely small compared to the volume of the atom. In a neutral atom, the number of electrons is always equal to the number of protons.

The notion of the atom may have originated in Babylon, Egypt or India, but the first concept remarkably similar to that of modern science was proposed by the 5th-century BC Greek philosopher Leucippus. He said that everything was composed of tiny particles so small that nothing smaller was conceivable. Democritus, a pupil of Leucippus, adopted and extended his teacher's ideas. He said that matter was composed of empty space and an infinite number of tiny invisible particles that were indivisible. He called them *atomos* or atoms. About a century later the great Greek philosopher Aristotle rejected Democritus' idea and said that that the matter was completely uniform and continuous. The influence of Aristotle was extraordinary. His concept of matter was basically wrong but it was accepted for 20 centuries until it was replaced by Dalton's atomic theory.

In 1808 English scientist John Dalton proposed an empirical model of the atom: atoms cannot be created, destroyed or divided. In 1903 another English scientist, J.J. Thomson, came up with the first experimental model: the atom is a solid sphere of positive charge and with negative electrons. In 1911 New Zealand-born English scientist Ernest Rutherford present a new model of the atom, known as the nuclear atom. The atom contains a core or nucleus of very high density and very concentrated charge. Most of the atom is empty space, with negative electrons, like the planets around the Sun, moving about the tiny central nucleus which contains positive protons. In 1920 Rutherford suggested that the nucleus also contained neutral particles neutrons.

Rutherford's model of the atom is still regarded as essentially correct. However, according to the modern atomic model an atom is thought to consist of two sections, a nucleus and a cloud of negative electric charge surrounding the nucleus. The nucleus contains protons and neutrons and the surrounding cloud electrons. The quantum model – a complex mathematical model – assumes electron location as a probability cloud.

The process of discovery still continues. In 2012, the discovery of the elementary particle, the Higgs boson, has added to our knowledge of the matter.

The number of protons present in an atom is known as its atomic number. Atoms with the same atomic number have the same chemical properties. The sum of the numbers of protons and neutrons in an atom is known as its mass number.

The information about the atom is usually given in the way shown below.

$$^{35}_{17}\text{Cl}$$

This gives us following information about the chlorine atom.

mass number (shown as superscript) = 35
atomic number (shown as subscript) = 17
number of protons = 17
number of electrons = 17
number of neutrons = 35 – 17 = 18

The term *isotope* is used to describe atoms of an element containing different numbers of neutrons in the nucleus. All the isotopes of one particular atom have the same atomic number because they have same number of protons, but have different mass numbers because they have different numbers of neutrons. For example chlorine gas consists of two isotopes, chlorine-35 and chlorine-37. The nucleus of each isotope contains 17 protons but different number of neutrons: 18 in one, 20 in the other.

5. Bacteria

Bacteria (singular: bacterium) are a diverse group of single-celled microscopic organisms with no nucleus. All organisms can be divided into *prokaryotes* (bacteria) and *eukaryotes* (all organisms except bacteria), the latter having a well-organised nucleus and chromosomes in each cell. Bacteria occur everywhere in the environment, even inside many living things. Most kinds of bacteria are harmless to humans; some, such as those that produce vitamins in the gut, are beneficial. A few kinds can cause disease. Bacteria display a vast range of characteristics. There are bacteria that can only survive in anaerobic conditions and bacteria that need oxygen to survive.

Depending upon how they feed, bacteria are broadly divided into two groups:

- Autotrophic bacteria obtain all or most of their energy by photosynthesis. They are 'self-feeders'. The group includes all plants and some bacteria such as cynobacteria (or blue-green algae; they are the most ancient forms of life on Earth, about 3.5 billion years old), green sulphur bacteria (they thrive in hot springs) and halophiles (as their name suggests these 'salt lovers' flourish in salty water)
- Heterotrophic bacteria are unable to manufacture their own food and are dependent on external sources. All animals and most bacteria are heterotrophic ('other feeders').

Life on Earth depends heavily upon the activity of bacteria. Carbon, in the form of dead plants and animals, would quickly deplete carbon dioxide from the atmosphere if not for the activity of bacteria. Bacteria are the most active decomposers of organic matter. When living organisms die, bacteria break down their bodies into simpler forms of matter that become part of the soil. Plants grow on these simpler forms of matter or nutrients.

Nitrogen fixation is the process of conversion of atmospheric nitrogen into forms that plants can use. Nitrogen fixation is extremely important for life on the planet because food production depends primarily on plants finding adequate amounts of nitrogen compounds in the soil. In nitrogen fixation, the atmospheric nitrogen is changed naturally into nitrogen compounds by (a) lightning and (nitrogen-fixing bacteria that live in the root nodules of plants such as peas, beans and clover). When plants and animals die and decompose, the proteins find their way back to the soil where putrefying bacteria break down the proteins into ammonia. Nitrifying bacteria in the soil then change the ammonia to nitrates. Denitrifying bacteria break down the nitrates and the free nitrogen is released into the atmosphere.

This process, known as the *nitrogen cycle*, keeps the amount of nitrogen in the atmosphere at a level of about 78 per cent. These days, nitrogen-containing fertilisers are a substantial source of nitrogen in the soil.

6. Big Bang

Once upon a time there was nothing – no planets, no Sun, not even stars in the sky. (Time did not exist either, so we can't even say 'once upon a time'.) Then there was this titanic explosion we now call the Big Bang, and our universe was born out of nothing. The Big Bang, which happened about 13.7 billion years ago, marked the beginning of time. Our galaxy was formed about a billion years ago; our planet about 4600 million years ago from a ring of gas and dust around the young Sun. Life began about 3500 million years ago.

The universe began when a single point of infinitely dense and infinitely hot matter exploded spontaneously. The debris of this explosion began to fly away from the explosion point and is still flying and will keep on flying indefinitely. All matter, energy, time and space were formed at this instant. After 10^{-43} seconds the universe inflated from the size of an atom to that of a grapefruit. After 10^{-23} seconds the universe was a superhot soup (10^{27}°C) of electrons, quarks and other particles.

After 380,000 years the universe cooled to about 4500°C and electrons and nuclei could combine to form the first hydrogen atoms. The universe now became transparent and photons were free to escape as gamma rays. As the universe continued cooling and expanding the wavelength of the radiation stretched. It changed from short wavelength gamma rays to longer wavelength X-rays, ultraviolet rays, visible light, and after 13.7 billion years into microwaves. This remnant radiation, usually referred to as the cosmic microwave background, fills the universe. This radiation was detected in 1965 and it has a temperature of – 270°C (3°C above absolute zero).

The idea of the big bang was suggested in 1927 by Belgian astronomer Georges Lemaître. He said that at some time in the remote past all the matter in the universe was concentrated at one point. The universe began when this 'primeval atom' exploded. This idea was further developed in 1948 by American physicist George Gamow who showed that as the universe began from a fireball, leftover warmth from this primeval fireball still filled the universe. This leftover radiation is now known as the cosmic microwave background radiation.

7. Biodiversity

Biodiversity – short for biological diversity – is diversity of life on Earth. Variety of plants and animals (*species diversity*) is only one aspect of biodiversity. Other aspects are *genetic diversity* and *ecological diversity*. Genetic diversity is the variation in genes that exist within a species (a species is a group of living organisms that can interbreed). Ecological biodiversity is the diversity of ecosystems, natural communities and habitats on our planet. Biodiversity can also be studied at a smaller scale: the word can be used to describe all the different kind of plants and animals found in a pond or a park.

The number of species per unit area – species richness – is a good measure of biological diversity. Biodiversity is considered to be high where there are the most species living in the same area. Because two overwhelmingly species-rich groups, arthropods (includes crustaceans, insects, arachnids and centipedes) and flowering plants are concentrated in tropical forests, the majority of the planet's species are believed to live there. Tropical forests have the most biodiversity. The world is destroying its rainforests at a record rate. That is one of the reasons why biodiversity is now at its lowest levels since the end of the Mesozoic era, about 70 million years ago.

Biodiversity is an issue of coexistence. Extinction – the disappearance of an entire species from our planet – is the extreme manifestation of a failure to coexist. Why do species become extinct? There are many causes:

- The failure to survive natural disasters.
- The failure to coexist as a result of competition, predation or disease.
- Failure to coexist with humans. Habitat destruction by humans is the biggest cause of biodiversity decline.

Extinction is not an unusual thing. – species continually disappear and new species

25

disappear. A species lasts from one to 10 million years and of all the species that ever lived on our planet 99.9 per cent are now extinct. But the current rate of extinction is alarming. Some environmentalists say that a species of animal or plant becomes extinct every 30 minutes.

Biodiversity is important for us because it provides all our food, most materials and medicines. It also helps in keeping air and water clean and soils fertile. Most medical discoveries are made because of research into plant and animal biology and genetics. Genetic diversity helps in preventing disease; it also helps species to adjust to their changing environment.

8. Bioethics

Bioethics is a philosophical discipline that deals with the ways advances in medicine and science touch upon our health, life and morality; and how these advances impact upon our society and the environment. The term 'bioethics' (from the Greek words *bios*, life; *ethos*, behaviour) was coined in 1927 by Fritz Jahr, a German philosopher. Jahr said that 'bioethics is a necessary moral attitude, conviction and conduct' and 'bioethics recognises and respects all life and living interactions in nature and culture'. Nowadays bioethics deals with issues as diverse as animal rights, cloning, euthanasia, gene therapy, organ transplants and stem cell research.

Here're brief comments on three bioethical issues.

Animal rights

Animal rights movement seeks to extend the basic moral principle of equal consideration of interest – which is now applied to all human beings – to animals as well. Most scientific laboratories now follow some kind of code of practice for the care and use of animals for scientific research. The guiding philosophy takes the idea that, unless there is evidence to the contrary, it must be assumed that animals experience pain in a manner similar to humans

Sex selection in IVF (*in vitro* fertilisation)

IVF is fertilisation of egg and sperm within a medium containing special nutritive substances in the laboratory. Six days after the fertilisation the embryo is transferred into the woman's uterus. The technique of preimplantation genetic diagnosis (PGD) allows recognition and elimination of an embryo with a genetic disorder or malformation and permits selection of sex. The International Bioethics Committee of UNESCO recommends that the use of PGD should be limited to medical diagnosis and sex selection for non-medical reasons should be considered to be unethical. Here the major bioethical issue is: Do parents have the right to decide the characteristics of their children?

'Brain death'

When our heart stops beating and there is no blood flowing to our brain, all brain activity ceases and we are declared clinically dead. Heart failure in itself does not constitute death. Pumping of blood is important, but it only supports the functions of the brain. Simultaneous monitoring of heart rate and brainwaves shows that it takes brainwaves 11 to 20 seconds to go flat when the heart stops pumping. 'Brain death' is arguably the most accurate biological representation of death. In 1968 a committee on medical ethics at the Harvard Medical School declared that a patient who is in coma and who was likely to die within 24 hours could be declared 'brain dead' for the purpose of using organs for transplantation purposes. Is brain death real death? The debate on this important ethical issue continues.

9. Biological Clock

We are all captives of our biological clocks – an internal timing system that regulates metabolism in all forms of life. Hundreds of cellular, physiological and behavioural patterns have been observed to follow a 24-hour cycle in humans. For this reason, the biological clock is also called circadian rhythm (from the Latin *circa diem*, about a day). Circadian rhythms are not related to popular, pseudoscientific concept of biorhythms. Body temperature is a good example of circadian rhythms. The body temperature is at its lowest in the early hours of morning and reaches its maximum in the late afternoon and early evening.

A temperature of 37 degrees Celsius is considered to be normal body temperature, but healthy individuals show a 24-hour cycle that varies from 35.5 to 38.5 degrees.

Jet lag and health problems associated with working in rotating shifts are caused largely by the body's battle against its circadian clock, the light-sensitive timepiece which also regulates sleep cycles. Defective clocks can trigger depression and sleep disorders.

The period of circadian rhythms is rarely exactly 24 hours, but varies from 23 to 25 hours. The human internal sleep–wake cycle is about 25 hours long. Because of the 25-hour sleep–wake cycle people are constantly advancing their sleep by an hour a day to conform to Earth's 24-hour schedule. But when people rotate shifts the change in sleep–wake cycle is too dramatic; the system becomes desynchronised and begins to 'free run' – to drift forward on its 25-hour cycle until it is back in phase.

The same desynchronisation is the cause of jet lag. Because the 25-hour sleep–wake cycle tends naturally to delay sleep, it is somewhat easier to adjust to work schedules that require us to stay up later than usual – a forward rotation.

The possession of the biological clock enables organisms to be in tune with their environment. Without it their survival in a hostile environment would not be possible.

Since all forms of life possess the biological clock and it is to their advantage to have the clock, it is likely that it has been developed during evolution. Researchers now believe that in vertebrates, for example, the clock system arose more than 450 million years ago.

If we do have a biological clock, where is it located? In mammals, including humans, the clock resides in the hypothalamus of the brain in a tiny collection of cells called the suprachiasmatic nucleus (SCN). The SCN lies close to the optic tract and is directly connected to the eyes. The SCN is but one part of the so-called circadian axis; the other two components are the pineal gland (which produces the hormone melatonin in the dark) and the retina.

In some people the excessive secretion of melatonin during the long dark nights (and dark) days of winter can trigger a condition known as seasonal affective disorder (SAD) or winter blues. This depressive state can be cured by exposure to suitable bright lights.

10. Biosphere

Biosphere is the zone on the Earth's surface containing life – the global sum of all ecosystems. The biosphere began when life began nearly 3500 million years ago. Every part of the planet – from polar caps to equator – supports some form of life. Life even exists deep beneath the Earth's surface. Biosphere interacts with other layers of the planet: lithosphere, hydrosphere and atmosphere. Energy from the Sun is essential for the existence of the biosphere. Extinction threatens the biosphere. Extinction is the disappearance of a species; that is, an entire species of animals or plants has died and can never return.

When the environment changes, species must adapt to the new environment to survive. Those unable to adapt become extinct. When the number of extinctions is very large compared to the number of extinctions that normally occur, they are called mass extinctions. The worst destruction of life in the Earth's history took place 245 million years ago. Scientists call this mass extinction the Great Dying because it nearly wiped out most of life on Earth. The death toll included 95 per cent of all living things in the oceans, 70 per cent of reptiles and amphibians, and 30 per cent of insects. What caused this spectacular extinction? The long line-up of suspects includes changes in world climate, sudden drop in sea levels, poisonous concentrations of carbon dioxide in the oceans, reduced oxygen and increased carbon dioxide in the atmosphere, decreasing supplies of nutrients in the oceans, huge volcanic eruptions, and a massive extraterrestrial object the size of Mount Everest that slammed into the Earth.

If Earth were a superorganism, as suggested by the *Gaia hypothesis* (Gaia was the name given by the ancient Greeks to their Earth goddess), the biosphere would have a better chance to survive the devastation caused by a mass extinction. If there were no life on Earth today, its atmosphere would be like that of Mars, with a lot of carbon dioxide and no oxygen. Life on Earth did not adapt to the conditions it found, but has helped to change the environment to make it hospitable; for example, nearly all the carbon dioxide, nitrogen and oxygen in the air today come from biological sources.

The Gaia hypothesis seeks to explain the interdependence of life and the physical world: the Earth's temperature and atmospheric composition are regulated by feedback between the living things and the physical world, and these feedback mechanisms have evolved with the biosphere over the past 3.5 billion years.

James Lovelock, a British scientist, first proposed the Gaia hypothesis in 1972. He said that life does not exist on Earth only because material conditions happen to be just right. Life on Earth defines the material conditions needed for its survival and makes sure that they stay there. The Earth's living matter, air, oceans and land surface form part of a giant system that is able to control temperature, the composition of the air and sea, the pH of the soil and other parameters so as to be optimum for survival of the biosphere. If the system is damaged dangerously, it can repair itself. The system seems to exhibit the behaviour of a single organism, even a living creature.

Lovelock sees Earth as a living organism of which we are a part; not the owner, nor the tenant, nor even a passenger on that obsolete metaphor, 'Spaceship Earth'. Any species that adversely affects the environment is doomed, but life goes on.

One of the predictions of the Gaia hypothesis is that there was never any life on Mars – or, if there was life in its early history, it should now exist in some form. The hypothesis maintains that life quickly gains control over the environment and regulates it. Mars was much warmer once than it is now, and if life existed on Mars in its early history, the slow cooling should not have wiped it all out. If the Gaia hypothesis is correct, then life should exist on Mars now.

11. Black Hole

Black holes are the end points in the life cycle of stars 10 to 15 times as massive as the Sun. Sometimes the crushing weight of a dying star squeezes it into a point of infinite density. At this point, known as singularity, mass has no volume and both space and time stop. The singularity is surrounded by an imaginary surface known as the event horizon, a kind of one-way spherical boundary. Nothing – not even light – can escape the event horizon. Matter falling into it is swallowed and disappears forever. That's why these regions of space-time are called black holes.

A black hole is a star that has stopped twinkling. But why?

In astronomers' jargon our Sun is a main sequence star. A main sequence star – and 90 per cent of stars are these – fuses hydrogen nuclei into helium nuclei at its centre. The Sun has lived 4600 million years as a stable star, and many billion years lie ahead. After consuming its hydrogen, the Sun will begin to expand. It will change into a type of star known as a giant, and will be about 100 times brighter than it is now.

After a few thousand years, the giant Sun will completely exhaust its supply of hydrogen and will shrink into a white dwarf – no larger than Earth, but so heavy that a teaspoonful of its matter would weigh thousands of kilograms. A white dwarf is so hot that it shines white-hot. Over billions of years, the white dwarf will turn black and cold. It will now be a dead star – a black dwarf.

A heavyweight star (10 to 15 times as heavy as the Sun) has a dramatic but brief life after becoming a supergiant. It spends its fuel so extravagantly that it collapses within a few million years. It then explodes as a supernova, which ejects an enormous amount of matter and even outshines the entire galaxy for a few days. The remaining matter forms a neutron star, only about 25 kilometres across, which contains tightly packed neutrons. These neutron stars do not glow, and are so heavy that even a pinhead of their matter would have a mass of a million tonnes.

If the crushing weight of a neutron star squeezes it into a point of zero volume and infinite density, it becomes a black hole.

The radius of a black hole is the radius of the event horizon surrounding it. This is called the Schwarzschild radius, after the German astronomer Karl Schwarzschild who in 1916 predicted the existence of black holes. A black hole's weird effects occur within 10 Schwarzschild radii of its centre. Beyond this rather limited distance, the only effect is through the black hole's normal gravitational pull. So, contrary to popular belief, a black hole is not like a cosmic vacuum cleaner that sucks in everything around it.

Not that long ago, black holes were in the realm of science fiction, but now there is convincing evidence for their existence. This evidence is still circumstantial – there is no way black holes can be observed directly. There are at least two species of massive black holes: smaller ones (a few times as massive as the Sun) that orbit normal stars; and their supermassive siblings (weighing many million Suns) which lurk in the centres of most galaxies. Our galaxy is believed to have a relatively small black hole that is as massive as 2.6 million Suns. A black hole with a mass 100 million times that of our Sun and a radius of 25 million kilometres squats at the centre of a galaxy 130 million light years away.

12. Carbon

Carbon is present in all plants and animals. Carbon has an extraordinary ability to form compounds with other elements. A carbon atom can form bonds with four other atoms. These atoms may be other carbon atoms or non-metal atoms, especially hydrogen, oxygen, nitrogen, sulphur and phosphorus. Most organic compounds are combination of carbon with one or more of these five atoms. Carbon-based molecules in terrestrial life cannot obtain liquid water essential for their wellbeing below freezing point, and they start breaking down above a few hundred degrees. This narrow range of temperatures makes them suitable for life on Earth only.

Silicon also has some life-giving properties of carbon: it can form long chains to which other elements bind, and it is abundant in the universe (sand is silicon dioxide), but not as much as carbon. These peculiarities of silicon have led some scientists to suggest that there might be silicon-based life on other planets. However, Swedish chemist Svante Arrhenius was totally against this speculation. He said in 1908: 'All organic beings in the whole universe should be related to one another, and should consist of cells which are built up of carbon, hydrogen, oxygen and nitrogen. The imagined evidence of living beings in other worlds in whose constitution carbon is replaced by silicon or titanium must be relegated to the realm of improbability.' The eminent American astronomer Carl Sagan, 'a carbon chauvinist' in his own words, said in 1976: 'Carbon compounds are not just more abundant but more stable. So while life on other planets would probably look like life on Earth because its internal biochemistry would be astoundingly different, I think it would be a based on carbon.'

While scientists root for carbon, science fiction writers relish the idea of silicon-based life. Even the gifted H.G. Wells was 'startled by the visions of silicon-aluminium organisms … wandering through an atmosphere of gaseous sulphur'. More recently, in 2267 in fact, *Star Trek*'s Commander Spock has encountered silicon-based life, the Horta, on the planet Janus VI.

The best literary essay on carbon is the last chapter titled 'Carbon' in renowned Italian author Primo Levi's *The Periodic Table* (1975), an extraordinary mixture of science and personal reminiscences. Named the best science book ever by the Royal Institution of Great Britain in 2006, *The Periodic Table* is considered one of the most important literary masterpieces of the 20th century. An excerpt:

'Carbon, in fact, is a singular element: it is the only element that can bind itself in long stable chains without a great expense of energy, and for life on earth (the only one we know so far) precisely long chains are required. Therefore, carbon is the key element of living substance ... If the elaboration of carbon were not a common daily occurrence, on the scale of billions of tons a week, wherever the green of a leaf appears, it would by full right deserves to be called a miracle.'

13. Cell

Cells are the basic unit of all living things except viruses. Cells vary in size, shape and function. There are two main types of cells: (1) Prokaryotic cells have no membrane-enclosed nuclei – bacteria, and archaea (single-celled microscopic microorganisms which live in extreme environments) fall in this group. (2) Eukaryotic cells have nuclei – all other organisms fall in this group. Most plant and animal cells vary in sizes from 1 to 100 micrometres. The number of cells in an average human body is estimated to be between 50 and 100 trillion, and about 90 per cent of these cells are bacteria.

Living cells were discovered by English scientist Robert Hooke, a contemporary of Newton. He named them so because they reminded him of tiny monks' rooms or cells in monasteries. Scottish scientist Robert Brown was the first to recognise a small body within cells as a regular feature of cells. He named this the nucleus (from the Latin for 'little nut').

A typical animal cell consists of an outer thin plasma membrane, which encloses the protoplasm. The protoplasm is divided into the nucleus and a viscous fluid called the cytoplasm

The nucleus is bounded by a double-membrane and contains a semi-fluid substance called the nucleoplasm, which is enclosed in a double-membrane system called nuclear envelope. The nucleus contains a small body called the nucleolus. RNA is made in it. The nucleus also contains the chromatin, which are masses of DNA and associated proteins. During cell division, chromatins become tightly coiled into chromosomes.

The cytoplasm also contains the Golgi body, or Golgi apparatus, which assemble and package proteins before they leave the cell. The rod-like organelle mitochondrion contains enzymes and it is here that the mist chemical reactions of respiration take place. A cell may have as many as thousand mitochondria.

Lysosomes are dark spherical bodies in the cytoplasm. They contain a host of enzymes which can break down complex compounds into simpler subunits. Endoplasmic reticulum is a membrane system in the cytoplasm. Attached to it are small particles called ribosomes, which contain RNA and protein.

Cells division and reproduction can occur in two ways:

- Mitosis produces two daughter cells that are identical to the parent cell; that is, each contains an identical copy of DNA. Daughter cells have exactly the same number of chromosomes as the parent cell. Mitosis allows plants and animals to grow and repair damaged tissue. It's also the basis of asexual reproduction in plants.
- Meiosis produces daughter cells that have half the total number of chromosomes as the parent cells. This process enables organism to reproduce sexually. In male humans, the body uses meiosis to create sperm cells; in females, it uses meiosis to create eggs.

As far as elements are concerned, cells consist mostly of carbon, hydrogen, oxygen and nitrogen, together with small quantities of sulphur and phosphorous. Plants and animals also require numerous other trace elements (small concentrations of elements present in a sample) such as sodium, potassium, magnesium, iron, zinc, calcium, chlorine, fluorine and copper. Of the 92 elements present in nature, only 21 play a role in life.

14. Chaos

Chaos theory is a new and exciting area of science that describes disorderly systems. The behaviour of a chaotic system is difficult to predict because there so many variable or unknown factors in the system. Chaos is a dynamic phenomenon. It occurs when the state of a system changes with time. Even simple systems can grow exponentially* with time, making long-term prediction of the future impossible. The behaviour of a dynamic system depends on its small initial conditions. In a chaotic system, even a small change can bring about major upheaval. Chaos helps scientists to understand the complexities of nature.

'Small differences in the initial conditions produce very great ones in the final phenomena. A small error in the former will produce an enormous error in the latter. Prediction becomes impossible.' Jules-Henri Poincaré, the renowned mathematician and philosopher of science, made this observation in 1908 in his book, *Science and Method*. Poincaré's observation received little attention from his contemporaries, but it has now earned him the title of the 'founder of chaos theory'.

In 1963 Edward Lorenz, an American meteorologist, was the first to study chaotic behaviour in nature when he developed a computer model to predict weather patterns. While working at the Massachusetts Institute of Technology, he developed a simple computer model to forecast changes in weather at a number of places. In one of his equations he used a rounded number; for example, 0.506127 became 0.506. He was surprised to see that his model now predicted quite different conditions. This suggested that even a small initial unpredictable condition such as a flapping butterfly could produce a larger global change in weather.

This is now called the 'butterfly effect': an action as small as a butterfly flapping its wings, say in Beijing, could bring about a snowstorm weeks later thousands of kilometres away in New York. Chaotic behaviour occurs in phenomena as diverse as the stock market, disease epidemic, population changes and the human heartbeat. Chaos theory, which touches all disciplines of science, can be used to examine the apparently random unpredictable features of the everyday world, such as the turbulent flow of water, traffic jams, the path of the winds and the build-up of clouds.

In 1975, American scientist Benoit Mandelbrot pioneered the mathematics fractals (a term he coined from the Latin *fract*, 'broken'). His fractals helped to picture the actions of chaos, rather than explain it.

Most patterns in nature aren't formed of simple geometric figures such as squares, triangles and circles, but of shapes that are jagged and broken up. Before Mandelbrot, mathematicians disdained describing such shapes mathematically. Classical geometry cannot describe the shape of a cloud, a mountain, a coastline or a tree. 'Clouds are not spheres' as Mandelbrot says, 'mountains are not cones, and bark is not smooth, nor does lightning travel in a straight line'.

Mandelbrot invented a new geometry of irregular and fragmented patterns around us. He called these beautifully complex patterns fractals. 'Small parts are the same as the big parts; that's the definition of fractal', says Mandelbrot. 'A cloud is made of billows upon billows that look like clouds. As you come closer to a cloud you don't get something smooth but irregularities at a smaller scale.' Ferns, cauliflowers, snowflakes, rivers, mountains and lightning – they all are fractals. Fractals can be described by simple mathematical equations that can be used to generate computer images.

Fractal geometry is now used to compress computer images; locate underground oil deposits; build dams; understand corrosion, acid rain, earthquakes and hurricanes; study global climate change; and even to model booms and busts of stock markets.

* In mathematics, an exponential is a function that varies with the power (exponent) of another quantity. In $y = a^x$, y varies exponentially with x, the exponent. When x increase, y grows rapidly. This is called exponential growth. A J-shaped curve, commonly known as J curve, shows exponential growth.

15. Chemical Bond

A chemical bond is a strong force of attraction linking two or more atoms allowing the formation of chemical substances. There are various types: (1) An ionic bond is the force of attraction between the positive and negative ions (as in sodium chloride). (2) A covalent bond is formed by sharing of electrons between two atoms (carbon dioxide). (3) A metallic bond is the electrostatic force of attraction present in metallic solids (alloys such as brass and steel). (4) A hydrogen bond is a powerful intermolecular force present in compounds containing nitrogen–hydrogen, oxygen–hydrogen or fluorine–hydrogen groups (water).

Here're further explanations of these four types of bonds:

Ionic bond

An understanding of ions is necessary before you can understand ionic bonds. An ion is simply an atom or group of atoms which carries a positive or negative electric charge, resulting from a loss or gain of valence electrons. (Valence electrons are the electrons in the outer shell of the atom – they are most reactive and take part in chemical reactions.)

An ionic bond is formed when a metallic element loses one or more electrons to a nonmetallic element. Ionic compounds are usually crystalline solids. Sodium chloride is a typical ionic compound. It consists of cube-shaped crystals which consist of regular arrangement of positive sodium ions and negative chlorine ions.

Covalent bond

Sharing of electrons between atoms can take place in three ways: (1) A bond in which one pair of electrons is shared between two atoms is called a single bond. (2) Sharing of two pairs of electrons constitute a double bond. (3) Three shared pairs form a double bond.

Sharing of four or more electrons is not possible. Most covalent compounds are liquids and gases.

Metallic bond

In metallic solids crystal lattice points are occupied by positive ions. These ions are formed when a metal atom loses one or more of its outer electrons. These free electrons still remain in the crystal and move about freely. Metallic solids are held together by attractive forces between positive ions and a negative 'sea' of free electrons. These free electrons are also responsible for most metals being good conductors of heat and electricity.

Hydrogen bond

The best example of a hydrogen bond is water molecule. It has a negative charge at one end and an internal positive charge at other end. The negative end of one water molecule has a strong attraction for the positive end of another water molecule. This attraction results in a hydrogen bond. Because of hydrogen bonding considerable energy is needed to force water molecules apart.

16. Chemical Reaction

A chemical reaction is a chemical change in which one or more chemical elements or compounds form new compounds. The starting substances are called reactants and the substances formed products. A chemical reaction obeys the law of conservation of mass which states that matter is neither created nor destroyed in a chemical reaction. This means that the total mass of the products of a reaction must be equal to the total mass of reactants. Most reactions are reversible; that is, the products can also react to form original reactants. A chemical equation is a symbolic expression of a chemical reaction.

Six basic types of reactions are:

Synthesis: A reaction in which two or more simple substances combine to form a complex substance. For example, when hydrogen gas combines with oxygen gas to form water:

$$2H + O_2 = 2H_2O$$

Decomposition: In such reactions one compound breaks down into two more substances. For example, the electrolysis of water to form hydrogen and oxygen:

$$2H_2O = 2H + O_2$$

Single replacement: A single element replaces another in a compound. For example, the reaction between sodium and water to make sodium hydroxide and hydrogen:

$$2Na + 2H_2O = 2NaOH + H_2$$

Double replacement: In this reaction two compounds exchange parts. For example, the reaction between silver nitrate and sodium chloride producing sodium chloride and silver nitrate:

$AgNO_3 + NaCl \Leftrightarrow NaCl + AgNO_3$

Oxidation–reduction (redox): Oxidation is the loss of one or more electrons by an atom or molecule. Reduction is the gain of one or more electrons by an atom or molecule. Oxidation and reduction are reactions taking place at the same time: every time a loss of electrons takes place in a part of the reaction, an equal gain of electrons must take place in the other part of the reaction. For example, in the following reaction, the magnesium atoms and the oxygen molecule are neutral but the magnesium oxide is a product of oxidation. The magnesium atom has lost two electrons and the oxygen atom has gained two electrons.

$2Mg + O_2 = 2Mg^{2+}O^{2-}$

Acid–base reaction: *See* Acid/Base

17. Climate

Climate is the average weather pattern of an area over a period of time. Weather pattern conditions include temperature, pressure, rainfall, wind velocity, humidity, cloud types and depths, and hours of sunshine. Major climate changes occur over long periods, but human activity is also affecting the world's climate. There astronomical cycles affect our climate: *precession* (the Earth's axis traces out a small circle in 26,000 year cycle), *tilt* (tilt of the Earth's axis changes in 41,000 year cycle) and *orbital wobble* (the Earth's orbit changes shapes in 100,000 to 400,000 year cycles – Milankovitch cycles – were responsible for the ice ages).

The ice ages or glacial periods are periods in the Earth's history when ice sheets covered large areas of land not usually covered by seasonal ice. There have been many ices ages, each of which lasted a few million years. The last ice age occurred in the Pleistocene (the geological epoch which lasted from about 2.5 million to 11,700 years ago), and was made up of four glacial periods. The term Ice Age usually refers to these Pleistocene glacial periods. In present times about 10 per cent of the Earth's surface is covered by ice, but during the Ice Age about 30 per cent of the land was covered by ice.

Climate zones

Most systems used for classifying world climates are based on a system introduced in 1900 by Russian climatologist Vladimir Köppen who classified world climate into five major types:

- A, tropical rainy: no month cooler than 18°C
- B, dry: evaporation exceeds rainfall
- C, mild humid: the coldest month less than 18°C but more than –3°C
- D, humid cool, coldest months below –3°C and the warmest month above

10°C

- E, polar: the warmest month below –10°C

Sunspots and climate

Sunspots are like freckles on the Sun's bright face. Wherever magnetic fields emerge from the Sun, they suppress the flow of the surrounding hot gases, creating relatively cool regions, which appear as dark patches in the Sun's shallow outer layer known as the photosphere. The number of visible sunspots varies in a regular cycle, known as the sunspot cycle, reaching a maximum about every 11 years. Near a 'solar minimum' there are only a few sunspots. During a 'solar maximum' there is a marked increase in the number of sunspots and solar flares, which are huge bursts of energy released from the region of sunspots.

Ever since the 11-year sunspot cycle was documented in 1843, scientists have been fascinated by the possibility that the cycle might influence the Earth's climate. Satellite measurements show that the solar output indeed rises and falls in synchrony with the sunspot cycle. The flickerings are weak – only 0.1 per cent change in solar energy – but they do not show any upward or downward trend. Even so, they do influence temperatures on Earth. This influence, however, is ten times smaller than the effect of greenhouse gases over the 11-year sunspot cycle.

Human impact on the climate

Most scientists now believe that human activity is undoubtedly influencing climate change, even if we do not know absolute answers to 'when' and 'how much'. Studies show that the atmospheric concentration of carbon dioxide is rising steadily from pre-industrial levels. Mountain glaciers and snow cover have declined on average in both hemispheres. Scientists' computer models predict that sea level rises will range from 18 to 59 centimetres by 2100; and there is extremely high probability that extremes such as heatwaves and heavy rain will become more frequent, and tropical cyclones will become more intense.

18. Cloning

Cloning is the creation of an exact genetic copy of a cell, tissue or organism. A clone is a perfect copy of the original; every bit of their DNAs is similar. There are three different types of cloning: (1) Gene or DNA cloning creates copies of genes or segments of genes. (2) Reproductive cloning creates copy of an existing or previously existing organism. (3) Therapeutic cloning or 'embryo cloning' creates embryonic stem cells; which can be used to generate virtually any specialised cells in the human body. So far scientists have cloned sheep, goats, cows, horses, pigs, cats and rabbits.

Cloned organisms are genetically identical individuals produced from the same parent by non-sexual reproduction. Frogs and other animals have been cloned since 1975, but in 1996 cloning of Dolly the sheep was the first successful experiment to clone a mammal. It showed that mammals can be cloned from DNA taken from adult tissues.

To clone the sheep, Scottish embryologist Ian Wilmut and his team took cells from the tissues of mammary glands of a mature sheep. They then took eggs from another sheep, removed their nuclei which contain DNA, and fused the nuclei with the mammary cells by passing electric pulses through them. The process replaced the DNA of the egg with the genetic material from the mammary tissue. The cloned eggs were placed in a culture dish where they grew into embryos. The researchers cloned 277 eggs, of which only 29 grew into embryos. These they transplanted into 13 ewes, acting as surrogate mothers. About five months later only one lamb was born. The lamb, which was named Dolly, had no father and its genes entirely came from the udder of a ewe. Dolly the cloned sheep died in 2003.

In 2000 the first patent for cloning was issued to Wilmut's team. The mammal cloning experiment has been repeated successfully on other species of mammals. These experiments show that cloning of humans is possible, but it has major theological, ethical, moral and social implications. Our society seems little prepared for the ethical, psychological and legal complications carbon copies of human would inevitably bring. Then there is the fanciful and fearsome possibility that the cloning technique may be misused by unscrupulous leaders to serve their own ends by creating Frankenstein-like monsters or Hitler-like children.

Cloning animals is not a perfect science. There are risks associated with it. First, there is a high failure rate: for every 1000 tries, only 30 cloned copies are made. Second, all animals closed so far have shown some defects during later development. Third, most animals cloned so far have died young or died mysteriously.

These risks boost the argument against human cloning. Most scientists say that there is a role for cloning – and that role is in treating humans – not creating humans.

Incidentally, Dolly the sheep died in 2003 at the age of 6. She was mother to six lambs, all bred the old-fashioned way – not by cloning.

19. DNA

DNA is a threadlike molecule present in all living cells. The diverse and complicated genetic information is stored along its length in a simple universal chemical code. Not only does it store genetic information, it passes this information on to the next generation. The molecule consists of a double helix of two strands coiled around each other (like a twisted ladder). When the strands are uncoiled, they produce two copies of the original. This unique structure explains how DNA stores genetic information and how it passes this information on to the next generation by making an identical copy of itself.

Each 'side of the ladder' in a DNA (deoxyribonucleic acid) molecule is made up of chains of alternating sugar and phosphate units. 'Rungs' are made from pairs of four chemical compounds called bases: adenine (A), thymine (T), cytosine (C) and guanine (G). The bases always pair in a specific manner: A pairs with T, and C pairs with G. Thus there are only four combinations: A–T, C–G, T–A and G–C. The genetic code is the sequence of bases along the length of DNA. This code determines the order in which amino acids are linked together to form proteins.

The DNA resides in the nucleus of the cell. It's not directly involved in the functioning of the cell. Rather, it instructs the machinery of the cell to make required proteins. These proteins, in turn, control all chemical processes in the cell. The DNA directs the cell to make a molecule known as RNA (ribonucleic acid). RNA is made up of the same bases as DNA except that the base U (uracil) replaces the base T. RNA serves as the blueprint for making proteins needed by the cell. The instructions on RNA are in the form of a code that consists of combinations of three bases available on DNA. Each combination represents an amino acid. Of the 64 combinations possible, 61 represent the 20 amino acids (as only 20 amino acids occur in the cells of all organisms alive today, in some cases several combinations refer to the same amino acid). The other three combinations act as 'full stops' in the coded information. The code can be written in either DNA triplets or the RNA copy of triplets. As an example, the triplet TTA in DNA (or CUU in RNA) instructs the cell to add the amino acid leucine.

The sequence of base pairs along the length of the strands is not the same in DNAs of different organisms. It is this difference in the sequence that makes one life form different from another. The human genome, the full DNA sequence of humans, has about 3 billion base pairs, which are wound into 24 distinct sausage-like bundles, or chromosomes. A gene is a segment of a chromosome. It is a length of DNA which has a complete code for one protein.

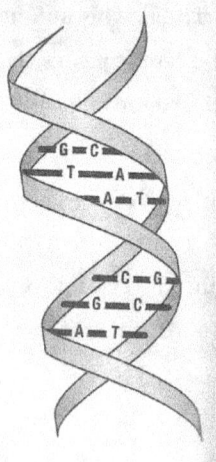

20. Earth

The third planet from the Sun and the fifth largest in the solar system, which lies 149,597,870 kilometres from the Sun, around which it revolves in 365.25 days, its orbit being slightly elliptical. It rotates on its axis from west to east every 24 hours with respect to the Sun (solar day), or every 23 hours 56 minutes 4 seconds with respect to the stars (sidereal day). It is flattened at the pole – its equatorial diameter is 12,756 kilometres, and polar diameter 12,713 kilometres. Its axis is tilted which gives rise to the seasons. It has one satellite, the Moon.

Earth is made up of layers, rather like an onion

The outer layer, the *crust*, has a thickness of about 35 kilometres in continental regions and about 10 kilometres under the oceans. Nine elements – oxygen, silicon, aluminium, iron, calcium, sodium, potassium, magnesium and hydrogen – make up 99 per cent of the crust.

Below the crust is the *mantle*, a thick shell of red-hot rock separating the core from the crust. Starting at an average depth of about 40 kilometres below the surface and continuing to a depth of about 2900 kilometres, it makes up about 82 per cent by volume of Earth, half its radium and 67 per cent of its mass. The mantle is mainly composed of magnesium-rich minerals, oxygen and silicon.

The very hot and dense central part of the Earth, the *core*, consists of a solid inner zone and an outer liquid zone. It contains mostly iron with a little nickel and cobalt. At the centre of the Earth the temperature is perhaps 3000°C, falling to 375°C at the mantle–crust boundary.

There is a layer of separation – a discontinuity – between the crust and the mantle. Known as *Mohorovičić discontinuity* (or simply Moho), it exists worldwide at an average depth of eight kilometres beneath the ocean basins and about 32 kilometres beneath the continents. There are two other boundaries: one between the mantle and the core, and the other between the liquid outer core and the solid inner core.

The outer solid part of the Earth, which takes in the crust and the upper mantle and is about 60 kilometres deep, is called *lithosphere*.

The part of the mantle from a depth of about 70 kilometres is called *asthenosphere* – it is weaker and hotter than lithosphere.

The part of the mantle below asthenosphere is called *mesosphere*.

All the water in or on the Earth's surface is called *hydrosphere*. About 70 per cent of the Earth's surface is covered with water, including the oceans, the seas, the lakes and rivers.

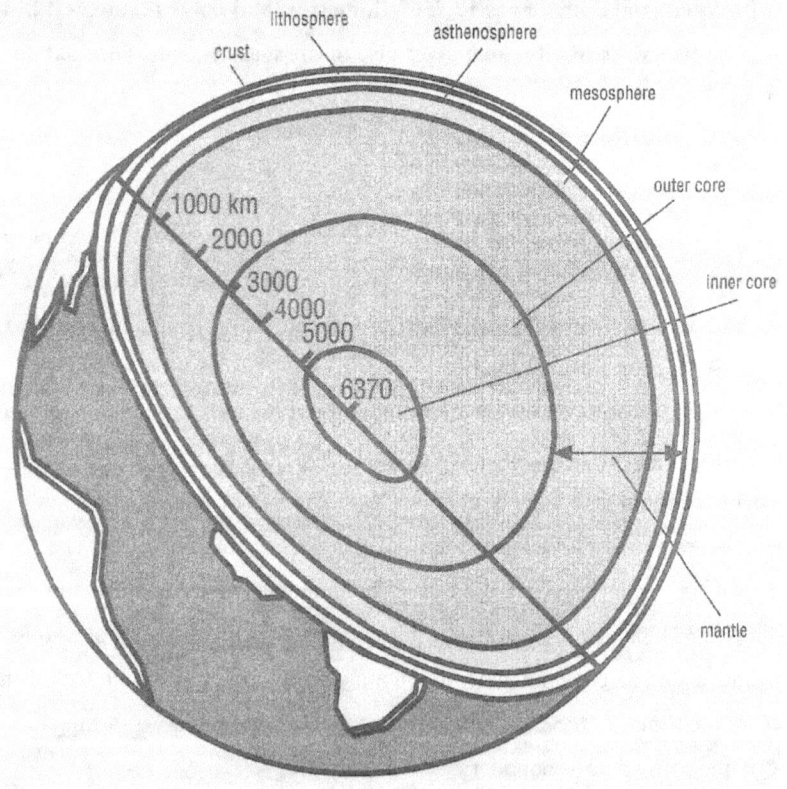

21. Ecosystem

A community is a group of different species living together in a particular habitat and interacting with each other. A habitat is the place, such as rainforests, where an organism or community of organisms live. Habitat + community = ecosystem. We describe an ecosystem as a community, such as a pond, a field, a forest, an ocean, even an aquarium. The global ecosystem includes all the nonliving or abiotic components (hydrosphere, lithosphere and atmosphere) and the living or biotic components (biosphere) of the Earth's environment. Ecology is the study of interrelationships of organisms to one another and to the environment.

An ecosystem includes producers, consumers and decomposers:

- *Producers* are organisms that convert sunlight into chemical energy. Typically, producers are green plants that produce energy-rich carbon compounds by photosynthesis.
- *Consumers* are organisms that feed on other organisms. Herbivores, which feed on plants, are primary consumers, and omnivores, which feed on both plants and animals, are secondary consumers.
- *Decomposers* are organisms that obtain energy from the chemical breakdown of dead organisms. Earthworms and most bacteria and fungi are decomposers. They are always the final stage in a food chain and return the nutrients to the environment in the inorganic state so that the nutrients can again be used by producers.

A *food chain* is a simple sequence of relationships showing the transfer of energy from producers to consumers to decomposers. For example, in a forest or garden we may see insect larvae feeding on leaves, birds eating the larvae and cats preying upon the birds. The food chain is: insect larvae – leaves – birds – cats –

decomposers. A complex food chain which branches out is called a *food web*: it represents the relationship between many different food chains.

The *ecological niche* is the role of an organism in the environment, its activities and relationships to the living and nonliving environment. An organism's niche depends on where it lives and on what it does. Habitat represents the 'address' of the organism, its niche indicates its 'profession'.

22. Electricity

Electricity is the flow of electric charge. Electric current is the motion of charged particles; in the case of conducting solids, the current-carrying particles are electrons. The conventional direction of the current is from positive to negative terminal in an electric circuit. (It was adopted before the discovery of the electron, and is still used to describe current flow.) Electric charge is the force of electric force. There are two kinds of charges: negative and positive. Unlike charges repel each other; like charge attract each other. Losing or gaining electrons (negative) with respect to protons (positive) makes a body charged.

Static electricity is known since the ancient times, but the 16th-century English scientist William Gilbert was the first to distinguish between magnetism and static electricity, which were considered similar phenomena. Amber produces static electricity when rubbed with a cloth. The Greek name for amber is *elektron*, from which Gilbert coined the Latin word *electricitas* for this property that amber displayed. The word was soon anglicised to 'electricity'.

The story of the discovery of current electricity is quite fascinating. There are various accounts of the story how twitching in the leg of a dead frog led Galvani, professor of anatomy at the University of Bologna, to his famous theory of 'animal electricity' in 1791. According to one version, the physician had recommended a soup of frogs' legs for his sick wife Lucia. He decided to cook it himself. He cut up some frogs and suspended their legs by copper hooks on an iron balcony of his house. He was astonished to note that the legs shook convulsively every time they chanced to touch the iron of the balcony. He repeated the experiment in his laboratory. He observed that leg muscles of a frog contracted when they touched with two different metals. He further tried a series of experiments and finally concluded that he had discovered 'animal electricity'. He expounded the wrong theory that the muscles of frogs were the source of electricity. Galvani had in fact made the first primitive battery in which electric current could be produced by two different metals in a suitable solution, but blundered when he tried to explain the source of the current.

When Alessandro Volta, professor of natural philosophy at the University of Padua, read of his fellow countryman Galvani's experiment, he tried it himself but disagreed with the theory of 'animal electricity'. He suspected that the current was produced somehow by two metals, copper and iron, not the frogs' legs. After experimenting for eight years, Volta found the answer in 1800 when he dipped discs of copper and zinc in a bowl of salt solution and obtained a continuous supply of electric current. His simple device was different from Leyden jars which stored static electricity produced by electrostatic machines and were capable of generating only a single discharge.

He reasoned that a much larger charge could be produced by stacking several discs separated by discs of flannel soaked in salt water. By attaching wires to each end of the 'pile' he successfully obtained a steady current. The 'voltaic pile' was the first battery in history – the first device to generate electric current.

In 1827 Georg Ohm, a German scientist, discovered the basic law of electricity, which simply states $V = IR$. Ohm was a teacher of mathematics and science in a high school when he started his experiments on current electricity. At that time current electricity was a qualitative study and there were no accurate ways to measure various electrical quantities. From his experiments he established that the flow of electricity in a conductor depends upon its length, its cross-sectional area, and the material of which it's composed. He provided accurate definitions of voltage across a conductor (V), the current flowing through it (I) and its resistance (R) and the relationship between them, which is now known as Ohm's law.

In the same year, French scientist André Marie Ampère showed that two current-carrying wires attract each other if their currents are in the same direction, but repel if their current are opposite; and the force of attraction or repulsion is directly proportional to the strength of the current and inversely proportional to the square of the distance between them. This observation, now known as Ampère's law, helped to develop a new area of study called electromagnetism.

In the SI, the unit of electric current is *ampere* (symbol A), voltage *volt* (V) and resistance *ohm* (Ω, the Greek uppercase letter omega). These unit names honour three great scientists who enhanced our knowledge of electricity.

23. Electromagnetic Radiation

Electromagnetic radiation is energy (in the form of waves) generated by changing electric and magnetic fields. The waves consist of electric and magnetic fields vibrating in harmony in directions at right angles to each other. They travel with the speed of light and differ from each other in their frequency. Electromagnetic spectrum is the range of energies over which electromagnetic radiation extends. As energy of photons is proportional to the frequency of radiation, the spectrum is usually arranged in order of increasing frequency (decreasing wavelength): radio waves, microwaves, infrared radiation, visible light, ultraviolet radiation, X-rays, gamma rays and cosmic rays.

Cosmic rays are intense radiation in which the nuclei of atoms have been accelerated to high energies within our galaxy and elsewhere in the universe. Their energy density is the same as that of starlight and the more uncommon particles have energies equal to the kinetic energy of a tennis ball moving at 100 kilometres per hour. The Earth's atmosphere is continually being bombarded by cosmic rays.

Generated in radioactive atoms and nuclear explosions, *gamma rays* are highly penetrating and can only be stopped by tens of centimetres of lead or metres of concrete. They can cause serious and permanent damage to living cells.

X-rays are given off when fast moving electrons lose energy very rapidly. Like gamma rays they are highly penetrating and in large doses can cause serious damage to living cells.

The Earth's ozone layer stops most of the harmful *ultraviolet radiation* (UV light) from the Sun. Prolonged exposure to ultraviolet radiation can cause skin cancer.

Infrared radiation is emitted by hot objects. The greenhouse effect is caused by the trapping of the Sun's infrared radiation by a blanket of so-called greenhouse gases around the Earth.

Microwaves can be used to heat food because they cause molecules within the food to vibrate. They are also used in radar and telecommunication.

Radio waves are used for transmission of audio and video signals. Radio wave frequencies are divided into seven equal bands: extremely low frequency (ELF), very low frequency (VLF), low frequency (LF), medium frequency (MF), high frequency (HF), very high frequency (VHF) and ultra high frequency (UHF).

Laser

Laser (*l*ight *a*mplification by *s*timulated *e*mission of *r*adiation) is a source of monochromatic and coherent light with wavelength in the visible, ultraviolet or infrared regions of the electromagnetic spectrum. Excited atoms emit light in the form of discrete packets of energy called photons. The emission may be spontaneous, in which case atoms emit randomly. But if the emission is stimulated by other photons, the emitted photon is joined to other photons in the wave at exactly the right instant and with the same frequency. The result: a tight parallel beam of monochromatic light.

Maser

A maser (*m*icrowave *a*mplification by *s*timulated *e*mission of *r*adiation) is a device for producing an intense source of microwave radiation by stimulated emission. Natural masers occur in the interstellar gas clouds. As different kinds of molecules emit masers with different frequency levels, astronomers can identify interstellar molecules by studying masers

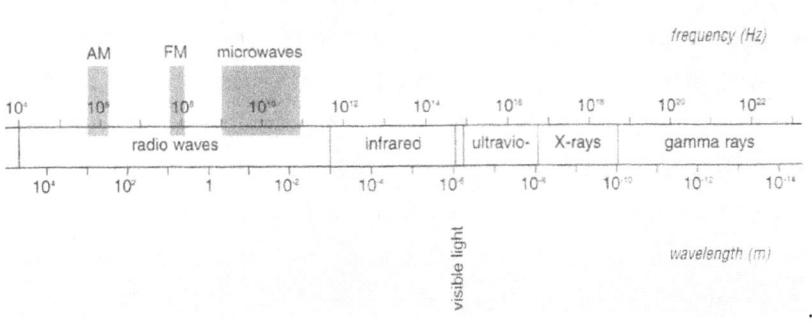

24. Electron

The electron is an elementary particle that exists in all atoms. It has a mass of 9.109x10^{-31} kilogram and a negative charge of 1.602x10^{-19} coulomb. Its mass is 1/1836 that of the proton; the charge is equal in magnitude but opposite. Two models of atomic structure are in use today. According to the Bohr model, the electrons in an atom revolve around the nucleus in definite orbits. According to quantum mechanical model, the location of the electrons around the nucleus cannot be precisely determined; the region where the electrons can probably be found is called the electron cloud.

In 1879 British physicist William Crookes performed an experiment using an evacuated glass tube with two metal electrodes sealed in opposite ends and connected to a high-voltage source. He noticed a stream of rays coming from the cathode, or negative electrode. He speculated that the cathodes rays, as the rays were called, consisted of tiny particles.

In the same year, another British physicist J.J. Thomson showed that cathode rays could be deflected by a magnetic or electric field. He concluded that cathode rays were streams of negatively charged particles; and these particles came from the atoms of the metal cathode. The Irish physicist George Stoney even suggested a name for these particles – the electron.

Thomson later measured the ratio of charge e to the mass m of the electron, which was refined in 1909 by American physicist Robert Millikan. His famous oil-drop experiment showed that the electron is the fundamental unit of electricity.

Millikan's apparatus simply consisted of a small box attached to a microscope. Surrounding the box were two brass plates. An atomiser introduced oil drops between the plates. By adjusting the voltage the charge on the plates could be changed until the drops were in mid-air. At this moment, the charge on the drop (upward electric force) equalled its weight (downward force of gravity). From this data, Millikan calculated the absolute charge on the electron, which is by convention called unit negative, –1, charge.

For their discoveries Thomson and Millikan were awarded Nobel Prize for physics in 1906 and 1923 respectively.

Two models of atomic structure in use today describe the location the electron in the atom in different ways.

The *Bohr model* says that the electrons in an atom revolve around the nucleus in definite orbits. The electrons in the orbits have certain amounts of energy, or electrons are said to be at certain energy levels. The electrons in the energy levels closest to the nucleus have the least energy, while the electrons in higher levels have more energy. Under normal conditions, the electrons in an atom are at the lowest energy levels. The atom is said to be in its ground state. If the atom picks up energy from an outside source it changes to an excited state, and the electrons are raised to higher energy levels.

The four main concepts of the *quantum mechanical model* are:

- The motion of an electron about an atom is not in a definite path.
- An electron's location can be given only in terms of probabilities.
- The location is described as a space orbital.
- The 'allowed' orbitals are regarded as being arranged in shells around the nucleus.

The region where the electrons can probably be found is called the electron cloud. The electron cloud can be considered in terms of a cloud of negative charge with the cloud being dense in regions of high electron probability and more diffuse in regions of low probability.

25. Element

Everything in the universe is made up of elements, which are simpler substances that cannot be split chemically into anything simpler. Elements are made up of atoms. Each element is represented by a symbol which stands for the element as well as for one atom of the element. The symbol of the element is usually the first letter of its name. Where the names of two or more elements begin with the same letter, two-letter symbols are used (caesium Cs, Calcium Ca, carbon C, Chlorine Cl, cobalt Co, copper Cu). Some symbols are based on Latin names (copper, *cuprum* Cu.)

One hundred and eighteen elements are known at the present time. Most of these are found in nature, but others have been synthesised. Elements heavier than 92 (uranium) – the heaviest element known to exist naturally in detectable amounts on earth – are all artificially produced, short-lived and radioactive.

The American-Italian nuclear physicist Enrico Fermi set the stage for creating new elements when, in 1933, he showed that the nucleus of most elements would absorb a neutron, changing the atom into a new element. In most cases this new element would be unstable and would decay into a different element. It occurred to Fermi that should be possible to prepare new elements by bombarding uranium nuclei with free neutrons. He tried the experiment in 1934 but was unsuccessful.

In 1940 American chemists Glenn T. Seaborg and Edwin McMillan produced and identified the first artificial element, neptunium, 93. In 1951 Seaborg and McMillan were awarded the Nobel Prize in chemistry for the discovery of elements 93 (neptunium) and 94 (plutonium). Seaborg is also the co-discoverer of many other elements. The name of element 106, seaborgium, honours him.

The discovery of the latest artificial element – element 118 – was announced in 2006. It is the heaviest atom ever produced and was created by Russian scientists at the Joint Institute of Nuclear Research at Dubna, near Moscow. Two other laboratories are at the forefront of discovering elements — GSI (the centre for heavy-ion research) at Darmstadt in Germany, and the Lawrence Berkeley Laboratory in California.

The International Union of Pure and Applied Chemistry (IUPAC) has yet to decide the names of elements 110 to 118. Until names are finalised, IUPAC recommends a temporary systematic nomenclature for new elements. Under this nomenclature, the digits of atomic numbers are translated into Latin-like syllables. Thus, element 110 would be called ununnilium, 111 unununium, 112 ununbium, 113 ununtrium, 114 ununquadium, 115 ununpentium, 116 ununhexium, 117 ununseptium, 118 ununoctium, and so on.

26. Elementary Particle

Elementary particles are the fundamental units of matter and energy. The best known of these units are electrons, protons, neutrons and photons. Each elementary particle has three characteristics: mass (some particles have zero mass), charge (every particle has a positive or negative or charge; some particles have zero charge) and spin (every particle spins somewhat like a top, except Higgs boson). Physicists believe that for every ordinary particle there exists a 'superpartner' with identical characteristics – except that its spin differs by half a unit. This is known as supersymmetry. There is still no direct evidence for the existence of supersymmetry.

According to the Standard Model of particle physics – a powerful theory that is central to the modern understanding of the nature of time, matter and the universe – all elementary particles fit into two categories: fermions and bosons.

Fermions (named after Italian-American physicist Enrico Fermi) are the particles of matter; they're only created in particle–antiparticle pairs. There are two classes of fermions: leptons and quarks. Both contain six particles: leptons – electron, electron neutrino, muon, muon neutrino, tau, tau neutrino; quarks – up quark, down quark, strange quark, charm quark, top quark, bottom quark. There are three generations of fermions. The first-generation particles make up the ordinary matter: protons (each an up-up-down quark triplet), neutrons (each an up-down-down quark triplet) and electrons. The second and third generation particles are produced in high-energy reactions and decay quickly into first-generation particles.

Bosons (named after Indian physicist Satyendra Nath Bose) are particles that transmit force by the exchange of an intermediate particle peculiar to that force. They are: gluons, *W* and *Z* bosons and Higgs boson.

The universe is held together by four types of fundamental forces:

- Gravity is the long-range force: it holds chair to the floor and planets in their orbits.
- Electromagnetic force is the attraction and repulsion between charged particles: it enables light bulbs to glow and lifts to rise.
- The strong force keeps atomic nuclei together: it binds together the protons and neutrons in an atomic nucleus.
- The weak force is also a kind of 'nuclear' force: it causes elementary particles to shoot out of the atomic nucleus during the nuclear decay of such radioactive elements as uranium.

The strong force is mediated by gluons, the electromagnetic force by photons, the weak force by W and Z bosons, and gravity by hypothetical particles called gravitons. The Standard Model does not include gravity.

The discovery of Higgs bosons in 2012 completes the Standard Model. The universe is filled with an invisible energy field. This field – known as the Higgs field – creates a drag on particles. If a particle moves through this field with little or no drag, it will have little or no mass. Alternatively, a particle interacting significantly with the field will have a higher mass. Higgs bosons suffuse the field and act as intermediaries between the Higgs field and other particles. When other particles attract Higgs bosons they acquire mass. It's the very reason matter exists in the universe.

27. Energy

Energy is a measure of a system's ability to do work. It is measured in joule (symbol J). The law of conservation of energy states that energy can neither be created nor destroyed, but may be transformed from one form to another. Energy sources are either renewable or nonrenewable. Fossils fuels (petroleum, coal and natural gas) are finite resources as it takes nature millions of years to produce them. They are nonrenewable energy sources. Nuclear energy, solar energy, hydroelectricity, geothermal energy, wind energy and biofuels are major renewable energy sources. Of these sources, nuclear nergy and biofuels are most controversial.

In physics, energy is classified into two forms:

- Kinetic energy is energy that an object possesses due to its motion. The kinetic energy of an object of mass m moving with a speed of v is $\frac{1}{2}mv^2$.

- Potential energy is energy that an object possesses due to its position shape or state. The most common example is gravitational potential energy. To raise an object to a height requires an amount of work, which is stored in the object as its potential energy. When you throw a ball straight up into the air, its kinetic energy transforms into its potential energy. When the ball falls back down again, its potential energy changes into kinetic energy. The gravitational potential energy of an object of mass m raised to height h is mgh, where g is the acceleration due to gravity.

Some comments on biofuels and nuclear energy

Most common *biofuels* are ethanol made from sugarcane or corn and biodiesel, a diesel-like fuel made from palm oil. Biofuels are renewable energy source because we can always more crops to turn into fuels. However, growing crops, manufacturing fertilisers and pesticides and processing plants into fuels also consumes significant amounts of energy.

Energy required for growing and processing plants comes from fossil fuels anyway. It raises the question whether biofuels actually provide more energy than is required to grow and process plants. Production of biofuels in industrialised countries is putting pressure on prices of crops such corn, sugarcane and palm oil, making them expensive for people in developing countries.

In the future, biofuels are likely to come from non-food crops such as grass, seaweed and ordinary plant matter.

Nuclear energy is relatively clean and safe but its advantages are debatable. One: Nuclear fuel is not a nonrenewable resource; supplies of high-grade uranium and other ores will run out well before the end of the century. Two: There is always a radiation risk, however minimal, from nuclear power stations to workers and the public in normal use and accidents. Three: There is no foolproof way yet for the storage of radioactive waste to safeguard the health of present and future generations.

28. Entropy

Entropy is a measure of disorder or randomness of a system. The more random or disordered a system is, the greater the entropy. For example, ice has low entropy. Its entropy increases when it melts into water, and increases much more when water turns into steam. Natural processes tend to move towards a state of greater disorder. The entropy is continually increasing in the universe. In a closed system, entropy must ultimately reach a maximum. Because the universe is a closed system, once all the energy in the universe converts into heat, there will be no energy available for work.

Once all the energy in the universe changes into heat, this will bring 'the heat death of the universe'. If entropy predicts that heat would end the universe, it also predicts that it will never freeze below absolute zero.

It's impossible to cool a substance to a temperature of absolute zero (−273.15ºC). Put simply, absolute zero is unattainable. We can approach it as closely as we wish, but we can't actually achieve it.

Entropy explains this paradox. When the temperature of a substance approaches absolute zero, its entropy approaches zero. Steam, for example, has a higher entropy than liquid water, as its molecules have a higher kinetic energy and move faster. Its entropy decreases when it condenses into liquid, and decreases much more when it freezes into ice. Its entropy would be zero if we continued cooling it to absolute zero, and all molecular motion (which is what constitutes heat) would stop completely. Atoms and molecules can't be idle; that's against the laws of physics – so this process is impossible.

29. Environment

Our physical and biological surroundings are called the environment. An organism's environment is the surroundings in which the organism lives and which influences the organism's distribution and survival. *Abiotic environment* **is the physical, or nonliving, components of an organism's surroundings. These include such features as light, temperature, atmospheric composition, water, the medium in which the organism lives, and the chemical environment.** *Biotic environment* **is the biological, or living, component of an organism's surroundings. It includes all living things. The environments may be regarded as natural or built; they may also be terrestrial, marine or atmospheric according to their natural features.**

Conservation is careful control and management of the planet's natural resources in order to preserve them as heritage. There are many reasons for protecting our environment:

- scientific (we need to learn about our environment before we destroy it)
- ethical (all species have right to live)
- genetic diversity (lost genes cannot be replaced)
- environmental diversity (diverse ecosystems are more stable)
- economic (unknown species could be used as food crops, medicinal plants etc.)
- aesthetic (plants and animals are beautiful)
- recreational (natural habitats have recreational value)

Issues affecting the environment include climate change, diminishing biodiversity, pollution, genetic engineering, depletion of resources, land degradation, overpopulation and nuclear technology.

Biome is a large natural area considered as a whole, with its particular climate, physical conditions, plants and animals. The Earth's land can be divided into several biomes: grasslands, deserts, tropical rain forests, coniferous forests, temperate forests, Arctic and Antarctic regions. The oceans can be regarded as a single biome – the marine biome. (*Biota* is the sum of all living organisms, including plants and animals, in an area.)

30. Enzyme

Enzymes are biological catalysts produced by living cells. Like ordinary catalysts, enzymes enable a reaction to take place yet themselves remain unchanged. A catalyst causes a chemical reaction to proceed, but it is itself not used up in the reaction. Enzymes are the largest and highly specialised class of proteins; and have a complex chemistry, controlling and regulating the biochemistry of the cell. The human body contains thousands of different kinds of enzymes. Each kind of enzyme catalyses a particular reaction. For example, an enzyme called amylase in our saliva increases the rate at which starch is converted into sugars.

Enzymes catalyse reaction in cells by forming a complex with the substrate, thereby increasing the probability that reaction will occur. (The substance an enzyme acts upon is called a substrate.) The complex then breaks apart to form the reaction products, freeing the enzyme to catalyse further reaction. For example:

sucrose (*substrate*) + sucrose (*enzyme*) = complex (*substrate-enzyme complex*) = glucose (*product*) + fructose (*product*) + sucrose (*enzyme*)

Note the above reaction takes place in presence of water.

Some laundry detergents contain enzymes. These enzymes catalyse the decomposition of proteins which make up the dirt of clothes. They are active only in wash which has a temperature of less than 60°C because higher temperatures destroy them. Enzymes in laundry detergents are generally safe as they are completely biodegradable and non-toxic to plants and animals in the environment.

Most enzymes work within a narrow temperature range. In our bodies enzymes work best at 37°C (body temperature). Enzymes also have an optimum pH. For example, the enzyme amylase works best at neutral or slightly acid pH.

31. Equilibrium

Equilibrium is the state of an object or system in which forces, influences, reactions etc. balance each other out. In physics, an object at rest or moving with uniform velocity is in static equilibrium – the sum of all forces acting on it is zero. An object at rest is in stable equilibrium if it is relatively difficult to move it from its original position (for example, a cone resting on its base on a table). An object at rest is in unstable equilibrium if a slight displacement causes it to move to a new position (cone placed on its tip).

Equilibrium (in physics)

Centre of gravity determines whether an object at rest is in stable or unstable equilibrium. Centre of gravity is a point where the whole weight of an object appears to be concentrated. It is the point at which the object will balance. (Centre of mass, on the other hand, is a point where the whole mass of an object appears to be concentrated. If an object is in a uniform gravitational field, its centre of mass is the same as centre of gravity.) A cone resting on its base on a table has a low centre of gravity making it stable. A cone resting on its tip has a high centre of gravity making it unstable as a slight force can shift its centre of gravity.

Chemical equilibrium

It is the state in a reversible reaction when the rate of forward reaction is equal to the rate of backward reaction.

Let us consider a very simple reversible reaction:

$$N_2 \text{ (gas)} + 3H_2 \text{ (gas)} = 2NH_3 \text{ (gas)}$$

At any one time, some nitrogen and hydrogen molecules are combining to form ammonia molecules. The rate of this process (the 'forward reaction') is proportional to the number of nitrogen and hydrogen molecules present. At the same, once ammonia molecules have been formed, they begin to break into nitrogen and hydrogen molecules. The rate of this process (the 'reverse reaction') should be proportional to the number of ammonia molecules present. As more ammonia molecules form, the rate of reverse reaction increases until it becomes equal to the rate of forward reaction. We say that the reaction has reached equilibrium. The equilibrium is dynamic; that is, the reactions are still occurring.

In a reversible reaction of the type:

$aA + bB = cC + dD$

the ratio of concentrations at equilibrium
$[C]^c[D]^d/[A]^a[B]^b$

is constant at a fixed temperature. The higher the value of equilibrium constant, the more complete the conversion of reactants into products.

The equilibrium constant of the above reaction between nitrogen and hydrogen to form ammonia is $[NH_3]^2/[N_2][H_2]^3$, and its value is 0.51 at a temperature of 943°C and a pressure of 101.325 kilopascal (760 millimetres of Hg).

A change in concentration, pressure or temperature can shift equilibrium. Catalysts do not affect equilibrium.

Nash equilibrium

Game theory is a mathematical method of analysing strategic behaviour – how people behave when placed in competitive situation. According to the theory all games have three things in common: rules, strategies and payoffs. Games include zero-sum games (each player benefits at the expense of others), non-zero-sum games (most real-world situations), cooperative games (people can make bargains) and games of complete information (rarely occur in the real world). The equilibrium of a game is called the Nash equilibrium, a solution that maximises everyone's benefit.

The theory has applications in economics, computer science, psychology, sociology, politics, warfare, evolution, the stock market and many other fields.

32. Error

The measurement of a physical quantity is always subject to some error. This error, or more precisely experimental error, contributes to the uncertainty of results. Experimental error is not a mistake, but a mathematical way of showing uncertainty in measurement. It's the difference between the result of the measurement and the true value of the quantity. Experimental error is usually caused by the limitation of measuring device, the process of measurement or the measuring environment. Simple mistakes such as incorrect use of an instrument or failure to read scale properly also cause experimental error, but they are easy to identify.

An *experimental error* is either random or systematic.

A *random error* is caused by the sensitivity of the measuring instrument or lack of skill in the experimenter. For example, a small ruler can cause more random error in measuring the length and width of a room than a long measuring tape. This ruler would have to be slid along the floor. A random error can be reduced by repeating a measurement a large number of times and averaging the result.

A *systematic error* is caused by a faulty measuring instrument (such as a fast-running stopwatch) or a fault in measuring procedure (such as mistimed action in starting or running a stopwatch). A systematic error cannot be corrected by repeated measurements but can be avoided by changing the conditions of measurement. However, it is possible that a new instrument or procedure may introduce a new set of systematic errors.

Accuracy is the closeness of a measurement to the accepted value for a specified quantity. Accuracy is expressed as either an absolute error or a relative error.

Absolute error is the difference between the measured value and the accepted value. Where a scale is has to be read, the actual size of absolute error depends on the sensitivity limit or *limit of reading*, the value of smallest division on the scale. Thus when using the scale we make a maximum possible error of half a scale division.

maximum absolute error = half the limit of reading of the measuring instrument

For example, when we measure the length of a book with a ruler marked in millimetres, the absolute error would be 0.5 mm. If the reading is 198 mm, it is evident that the length must lie between 197.5 mm and 198.5 mm, i.e. the length can be said to be 198 ±0.5 mm.

Relative error is the ratio of the absolute error and the accepted value. If the relative error is expressed as a percentage, it is called a *percentage error*. The percentage error in the above example would be $(0.5 \div 198) \times 100 = 0.25\%$.

33. Evolution

All present-day species have evolved from simpler forms of life through a process of natural selection. Organisms have changed over time, and the one living today are different from those that lived in the past. Furthermore, organisms that once lived are now extinct. Charles Darwin presented this theory of evolution in his monumental book, *On the Origin of Species*, published in 1859. Advances in modern biology, especially in knowledge of DNA, have enriched the theory of evolution. The modern view of evolution is still based on the Darwinian foundation: evolution through natural selection is opportunistic and it takes place steadily.

Darwin did not discuss the evolution of humans in *On the Origin of Species*. In a subsequent book, *The Descent of Man*, published in 1871, he presented his idea that humans evolved from apes. In this book, he set out reasons for thinking that humans had developed as a result of similar events to those which led to the evolution of other organisms.

Later fossil and molecular studies show that modern humans evolved from primates. The African apes are shown by anatomical and molecular comparison to be our closest living relatives. It is not known from fossil records when the human and African ape lines separated, but the molecular clock studies put it at 5 million years ago. There are no known hominid (human and humanlike) fossils from this period. The oldest known hominid fossils date from 3.5 million years ago, from Africa.

In modern form the theory of evolution includes the following ideas:

- Members of a species vary in form and behaviour and some of this variation has an inherited basis.
- Every species produces far more offspring than the environment can support.

- Some individuals are better adapted for survival in a given environment than others. This is called the 'survival of the fittest'. This means there are variations within each population gene pool and individuals with most favourable variations stand a better chance of survival.

- The favourable characteristics show up among more individuals of the next generation.

- There is thus a 'natural selection' for those individuals whose variations make them better adapted for survival and reproduction.

- The natural selection of strains of organisms favours the evolution of new species, through better 'adaptation' to their environments, as a consequence of genetic change or mutation.

34. Force

Put simply, a force is a quantitative description of a push or pull between two objects. In physics, a force is defined as a physical quantity that can change the speed or direction of motion of a body. A force can result from the contact or non-contact of the two interacting objects. Non-contact forces – or action-at-a-distance forces – are gravity, electric and magnetic forces. Friction is the most common contact force: it opposes the relative motion of two surfaces in contact. For example, when you push a book sitting on a table, an equal force of static friction opposes the motion.

In the above example, the static friction force reaches a maximum limiting force just before the book begins to slide. Once motion has started, kinetic friction is less than the limiting friction and acts opposite to the direction of the motion.

As mentioned above, a force changes the speed or direction of motion of a body. The change in speed or direction is acceleration. Thus force causes acceleration (a) of a mass (m) and is given by $F = ma$. The units of force is newton (symbol N).

Viscosity

In liquids and gases, viscosity is the frictional force that resists the flow of a liquid or gas. Viscosity is caused by frictional forces within itself – between different layers of the fluid as they move with different internal velocities. Viscous forces tend to slow down the faster-moving layers and to increase the velocity of the slower-moving layers. Different fluids have different viscosity: oil is more viscous than water; grease is more viscous than petrol; liquids in general are much more viscous than gases.

Centripetal force

It is the force required to keep an object moving in a circle. It always acts towards the centre of the circle and its magnitude, F, can be calculated by using Newton's second law of motion: $F = ma = mv^2/r = mr\omega^2$, where m is mass, v velocity, ω angular velocity and r radius. (Angular velocity, denoted by the Greek lowercase letter ω, is the time rate change of angular displacement.)

When you whirl a ball on a string you feel force pulling outwards on your hand. This experience gives rise to a common misconception that an outward force – centrifugal force – acts on an object moving in a circle. Essentially, the string is in a state of tension and pulls at both its ends; it pulls your hand outwards and the ball inwards. There is only one force – centripetal force – on the ball and this force pulls it inwards. Centrifugal force does *not* act on the ball. Centrifugal force is a myth.

Hookes's law

When an external force is applied to a solid its tendency is to maintain its size and shape. The internal forces created within the object as a result of applied force are stresses; the force causing the deformation are strains. Stress is force per unit area. Strain is the ratio of the change in length to the original length.

Within the limits of elasticity, the extension (strain) of an elastic material is proportional to the applied stretching force (stress). This is Hooke's law, discovered by English scientist Robert Hooke in 1665. The law applies to all kinds of materials, from rubber balls to steel springs. The helps define the limits of elasticity.

35. Fossil

Fossils are the evidence of past living things preserved mostly in rocks but sometimes in amber, ice, peat or tar. They are the keys that help palaeontologists (scientists who study fossils) to unlock the mysteries of time. Fossils can vary in size – from a large dinosaur skeleton to a tiny plant that can only be seen under a microscope. Some of the most perfect fossils are those of insects trapped in amber, which forms when resin or sap from coniferous trees hardens. As the saps oozes out of the tree it traps air, and occasionally, insects and other small animals.

Most abundant rocks on the Earth's surface are sedimentary rocks, covering two-thirds of the surface. They are formed from sediments – grains of fine sand or silt deposited by wind, water or ice. Remains of animals and plants are also buried with these sediments. As a layer of sediment is buried deeper by more sediment, it slowly turns into a sedimentary rock. Thus each distinctive layer contains fossils of living things that were common during the period the layer was formed. If you cut straight through a sedimentary rock the layers appear like the pages of a book viewed end-on.

Fossils of soft tissues are not common. When animals and plants die, their softer tissues begin to decay, leaving only hard parts such as shells, bones, teeth or wood to be preserved.

Soft tissues provide more information about the organisms than bones, teeth or shells do. Some of the most perfect fossils are those of insects trapped in amber, which form when resin or sap from coniferous trees hardens. As the yellow sap oozes out of the trees, it traps air and occasionally insects and other small creatures like frogs. From the trapped air bubbles, scientists can also determine the amount of various gases in the ancient atmosphere.

As animal remains are unlikely to be left undisturbed by scavenging animals, it is unusual to find full skeletons of large animals such as dinosaurs unless they are caught in a mudslide or buried in a sandstorm or volcanic ash.

A large animal like dinosaur has only one skeleton, but it can leave thousands of footprints. Fossil footprints are formed when the dent in the sediment is filled by sediment of contrasting texture. They can give us a host of information. For example, foot prints of many dinosaurs appearing together suggest they were social animals like elephants and lived in herds.

When studying the history of the earth scientists don't measure time in years or centuries – they use a scale known as the geological timescale. The major unit of geological time is called an era, which is in turn divided into periods.

The earth's rocks are like a pile of old newspapers in your home: the younger layers lie on top of the older layers. Each layer has a definite group of fossils of animals and plants that lived during the time the rock was formed.

The geological timescale is based on particular fossils of each layer of rock. Each period has a particular group of fossils that are unique to that time.

The geological periods and the major events that happened during each period are shown in the following table.

Era	Period	Beginning of the period (millions of years ago)	Major events
Palaeozoic	Precambrian	4600	First life appears about 3500 million years ago
	Cambrian	540	No life on land, but in the sea trilobites, corals, sponges, shellfish and jellyfish
	Ordovician	500	First fishes (first vertebrates, animals with backbones); abundant marine algae
	Silurian	435	Beginning of plants on land
	Devonian	410	The age of fishes; first amphibians and insects
	Carboniferous	355	First reptiles; coal forests; ferns and conifers
	Permian	295	Land reptiles
Mesozoic	Triassic	250	First dinosaurs; land dominated by non-flowering plants
	Jurassic	203	The age of dinosaurs; first mammals and birds; conifers; first flowering plants
	Cretaceous	144	End of dinosaurs; earliest birds; flowering plants increase
Cenozoic	Tertiary	65	The age of mammals: many kinds of mammals evolve, including horses, elephants and apes; coniferous forests and grasslands
	Quaternary	2	Ice ages and interglacial periods until 10,000 years ago\n\nFirst humans appear

Fossils are rare in rocks older than 540 million years (those formed before the Cambrian) so nearly 4000 million years of the Earth's history are taken together and are called the Precambrian. The oldest rocks found on the Earth – in the Northwest Territories of Canada – are nearly 4000 million years old.

36. Galaxy

Most of the material we see in the universe consists of stars and interstellar gas and dust collected together in galaxies, on of which is the Milky Way. There are 150 billion (and still counting) galaxies in the presently observable universe and each contains billions of stars. Their diameters range from several thousand to one hundred thousand light years. But we can only know of galaxies that are within a certain radius, known as the Hubble radius, as galaxies larger than that radius will be travelling with the speed of light. The Hubble radius is about 13 billion light years.

Most galaxies are classified into three basic groups according to the stars they contain:

- Elliptical galaxies contain relatively old stars distributed in a space shaped like a football and little or no interstellar gas or dust.
- Spiral galaxies are those in which elliptical-like central bulge is accompanied by a flat rotating disc with spiral arms wherein new stars are still being born out of clouds of gas and dust. Some spiral galaxies have a bar-like structure, extending from the central bulge where the spiral arms begin. The Milky Way is a spiral galaxy.
- Irregular galaxies have a disc rich in gas and dust clouds and young stars but little spiral structure and no bulge.

Our galaxy, the Milky Way, is 13.6-billion-year-old spiral disc of about 250 billion stars and huge clouds of gas and dust. This pancake-like disc is more than 100,000 light years wide (a light year is the distance that light travels in one year – about 9,500 billion kilometres). The spiral arms harbour young stars and are therefore brighter than the regions in-between. A 27,000 light years long, skinny bar of billions of relatively old, dim red stars cuts across the heart of the galaxy.

There is no dark syrupy chocolate at the centre of this Milky Way bar, but a dark fiery monster: a supermassive black hole, an infinitely dense point in space-time, sucking in everything around it. Every twinkling star we can see with the naked eye is part of our galaxy. Our solar system is situated in a position about 30,000 light years away from the centre of the Milky Way.

37. Gene

A gene is a segment of DNA chain that contains codes for the production of a particular protein or for a particular function. Genes are basic units of heredity and are located in chromosomes. Each of the 100 trillion cells in the human body carries in its nucleus 23 pairs of chromosomes (46 in all), except sperm and egg which contain only a single copy of each chromosome (23 in each). A special pair of chromosomes determines the sex of a human. Cells of women have two X chromosomes, whereas those of men have one X and one Y chromosome.

Scientists have now mapped all 22,000 genes in the human genome. These genes provide instructions for building the body's cells. However, these genes make up only 2 per cent of the human genome, the full DNA sequence of humans. Until recently, the other 98 per cent DNA was described as the 'junk' DNA as this DNA didn't seem to make proteins or have any other function. Now scientists are finding that most of the so-called 'junk' DNA is involved in some kind of biochemical activity. It's not junk after all.

The mapping of genes is helping scientists to find genes associated with dozens of genetic conditions and in developing thousands of new drugs for previously untreatable diseases.

Genetic engineering is a cutting-and-splicing technique in which a gene is cut from one kind of organism and pasted into another. The oldest genetic engineering technique is known as recombinant DNA technology. In this technique a DNA strand is cut with an enzyme. The fragment is then placed in a solution containing plasmids, small circular pieces of genetic material found in bacteria. The DNA fragment combines with a plasmid to form a new gene. The new gene, when placed in a solution containing normal bacterium, enters the bacterium. The bacterium then treats the new gene as its own and begins to produce the protein according to the new gene code. If, for example, the new code is for producing insulin, the bacterium will start producing insulin. In fact, in 1982 human insulin, produced by bacteria that had received the appropriate human gene, became the first product of the genetic revolution to reach the medical marketplace.

Another application of this revolution is gene therapy, in which a gene that's missing or defective is replaced with a correct one. Not a practised reality yet, but that's how one of the techniques of gene therapy works: the gene from a healthy person is spliced into the genes of an engineered harmless virus. The virus infects a cell, say a defective bone-marrow cell, taken from the patient and inserts the normal gene into the cell's DNA. The corrected cells are then injected back into the patient and multiply into healthy cells which produce the protein expressed by the gene.

38. Genetically Modified

Organism

A genetically modified organism is an organism whose genes have been modified using genetic engineering techniques (also known as genetic modification, genetic manipulation or recombinant DNA technology). In genetic engineering, a gene is cut from one kind of organism and pasted into another. This technology is now so advanced that virtually any gene can be inserted into any organism. This cut and paste can even take place between plants and animals. Genetically modified does not necessarily mean that a gene from another organism has to be used to create the genetically modified organism; organism's own genes can also be changed.

Ever since the technology became available scientists have dreamed of making new, improved plants and animals, and they have shown no lack of imagination: their experiments range from chickens that grow faster on less food to kidney beans that cause no flatulence.

The first plant product to leave the laboratory bench top and reach the dining table is a tomato called Flavr Savr. Endorsed in 1994 by the Food and Drug Administration as the first genetically modified food to be sold in the US, Flavr Savr was produced by suppressing a gene associated with an enzyme that makes the tomato rot. By reversing the effect, scientists ensured that the tomato stays fresh longer.

Genetic engineers can arm plants with defences against herbicides, insects and viruses, and produce nutritious and higher-yield crops. They also hope to use genetically engineered plants and plant viruses to produce vaccines against human diseases, ranging from tooth decay to AIDS.

But there is the flip side: some worry about potential health effects or damage to the environment; others worry about disease-resistant new plants passing their new genes to weeds, giving them an additional advantage to survive. Virus-resistant crops could potentially introduce new virulent viruses. Or, pollen from a genetically modified crop may carry to a conventional crop and this modifying its genes. Whatever the arguments may be in favour or against transgenic plants we are not likely to be overwhelmed by plants as in John Wyndham's classic novel, *The Day of Triffids*.

The biology of genetic engineering

Genetic engineering is a popular name for recombinant DNA technology. Recombinant DNA is a cutting-and-splicing technique by which DNA molecules are recombined or reassembled to produce different kinds of DNA molecules.

This can be achieved as follows: DNA isolated from an organism, say a mouse, is treated with restriction enzymes which can cut DNA at a specific point. The two fragments of DNA are each left with a terminal unpaired strand – a so-called 'sticky' end. The fragments of mouse DNA are then mixed in a solution with plasmid (small pieces of DNA which can duplicate independently) removed from a bacterium, say *E. coli*. The solution also contains a sealing enzyme, called DNA ligase, which cements mouse DNA into place in the opening of *E. coli* plasmid. The new plasmid when placed in a solution containing normal *E. coli* enters the bacterium. The bacterium then treats the new DNA as its own.

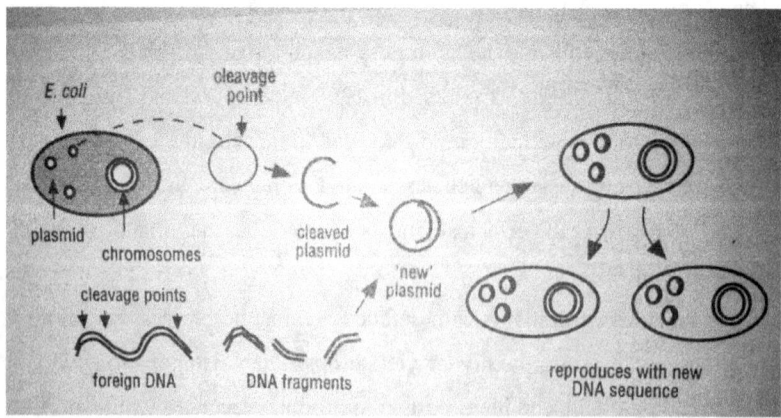

39. Gravity

Gravity is the dominant force in the universe. It's the force of attraction between all matter and a long-range force. It acts at a distance In 1687 Newton showed in his book *Principia* that a single universal force (a) keeps the planets in their orbit around the Sun, (b) holds the moons in their orbits, (c) causes objects to fall, (d) holds objects on the Earth, and (e) causes tides. But Newton failed to show how gravity works. (Gravity is the same thing as gravitation. The word gravity is particularly used for the attraction of the Earth for other objects.)

In *Principia* Newton published his famous law of universal gravitation: two bodies in the universe attracts each other with a force which is directly proportional to the product of their masses and inversely proportional to the square of the distance between them. It may be expressed by the equation $F = Gm_1m_2/r^2$, where m_1 and m_2 are the masses of two particles, r the distance between them and G the *universal gravitational constant*. The value of G is 6.672×10^{-11} N m^2/ kg^2.

Gravitational field is the region of space in which a gravitational force is exerted on a mass. The Earth's gravitational field always points to the centre of the Earth and its magnitude decreases as one moves away from the Earth.

Gravitational field strength is the pull of gravity on a mass of one kilogram. The Earth's gravitational field strength is 9.8 N/ kg. Other planets have different Earth's gravitational field strength. For example, Jupiter's gravitational field strength is 25 N/kg. As weight = mass × gravitational field strength, your weight on Jupiter will increase by 25 times. However, your mass will be the same. The Moon's gravitational field strength is merely 1.6 N/kg. That's why astronauts can jump high into the Moon's atmosphere.

While sitting in his garden, Newton noticed an apple drop from a tree. This set him to wondering why it should fall straight down to the ground. He came to the conclusion that the apple fell downward because some force was pulling it. This chance observation led him to his great theory of gravitation. Probably apocryphal, this is perhaps the most popular anecdote in science. The anecdote does not tell us whether the apple hit Newton on his head or not, but it is well known that it hit upon a scientific idea that solved many mysteries of the universe. In 1905, without the benefit of a falling apple, Einstein, gave a new meaning to Newton's force of gravitation.

40. Greenhouse Effect

Greenhouse effect is the warming of Earth as radiation is trapped in the atmosphere. Water vapour, carbon dioxide, methane and nitrous oxide make up less than 1 per cent of the Earth's atmosphere, but they play an important role in keeping our planet at a constant temperature. These natural gases – and some human-made gases such as chlorofluorocarbons – allow the sunlight to enter freely, but absorb heat radiated from the Earth's surface. Nature's thermal blanket keeps our planet at an ideal temperature by balancing the amount of heat received from the Sun with the amount of heat lost from the surface.

In 1896, Swedish chemist Svante Arrhenius was the first to recognise the greenhouse effect, the cause of global warming. He suggested that the burning of fossil fuels would increase carbon dioxide in the atmosphere, which would bring about significant changes to our climate – but Arrhenius's warning was ignored by the scientists of the day.

Burning of fossil fuels and destruction of forests continues to push the concentration of carbon dioxide in the atmosphere. The concentration of nitrous oxide, ozone and other harmful gases is also increasing. As these 'greenhouse gases' build up in the atmosphere and absorb the heat radiated from the Earth's surface, less heat escapes to space. Thus the greenhouse gase create a thermal blanket around the globe, with the potential to modify global climates, melt glaciers and raise sea levels.

Carbon dioxide is the most abundant of greenhouse gases. The main sources of carbon dioxide are from burning coal, oil and gas. At present the Earth's lower atmosphere (troposphere) contains about 390 parts per million (ppm) carbon dioxide – a clear increase from 315 ppm in 1957 when scientists internationally began precise and continuous measurements. Studies of air bubbles trapped in Antarctic ice core shows that in the 1850s the natural concentration of carbon dioxide was about 270 ppm, and roughly the same concentration persisted throughout the past 10,000 years.

Molecule for molecule methane traps 25 times more heat than carbon dioxide. The major source of atmospheric methane are anaerobic bacteria in animal waste, cattle and human flatulence, rice paddies, garbage tips, coal mining and leaks from natural gas distribution. Each cow, camel, sheep, goat, buffalo or other ruminant animals belch between 200 and 300 litre of methane into air every day. Termite, or white ants, which convert the carbon in wood into methane are another significant source of atmospheric methane. At present the troposphere contains about 1870 parts per billion (ppb) methane, a substantial increase from pre-1750 estimate of 700 ppb.

Molecule for molecule, nitrous oxide is 140 times more effective at global warming than carbon dioxide. The major sources of atmospheric nitrous oxide are the burning of forests and the use of nitrogen fertilisers. At present the troposphere contains about 325 ppb nitrous oxide, a clear increase from pre-1750 estimate of 270 ppb.

Ozone is found in the upper atmosphere (stratosphere) where it absorbs harmful ultraviolet rays; bit if found in the troposphere it is considered a pollutant and contributes to global warming. Molecule for molecule, ozone is 430 times more effective at global warming than carbon dioxide. At present the troposphere contains about 34 ppb nitrous oxide, an increase from pre-1750 estimate of 33 ppb.

Although the amount of chlorofluorocarbons in the atmosphere is very small, molecule for molecule they are thousands of times more effective at global warming than carbon dioxide. Recent measurements show an increase in the amount of these gases in the troposphere from pre-1750 estimates of zero. The manufacture of chlorofluorocarbons used as refrigerants, aerosol propellants and solvents and is being phased out.

41. Heat

Heat is a measure of energy – energy contained in a body in the form of kinetic energy of atomic or molecular rotation or vibration. Heat transfers from one body to another as result of difference in temperature. If two bodies at different temperatures are brought together, energy flows from the hotter body to the colder. This transfer of energy usually results in an increase in the temperature of the colder body and a decrease in the temperature of the hotter body. A substance may also absorb heat without an increase in temperature by changing from one physical state to another.

Heat it transferred from one place or body to another in three different ways:

- Conduction is the transfer of heat from high-kinetic-energy molecules to neighbouring lower-kinetic-energy molecules through successive collisions.
- Convection is the transfer of heat in a liquid or gas by the mass movement of high-kinetic-energy molecules from a high-temperature region to a low-temperature region.
- Radiation is the emission of rays, wave motion or particles from a source. In this way of heat transfer energy is transferred by electromagnetic waves at the speed of light. Unlike conduction and convection, radiation does not require the presence of matter.

Thermodynamics ('thermo' refers to heat and 'dynamics' to work) is the study of processes in which energy is transferred as heat or work. The three laws of thermodynamics are:

- First law of thermodynamics: Energy is neither created nor conserved destroyed, but may only be changed from one form to another. This law is one of the great laws of physics. It's simply a restatement of the law of conservation of energy.

- First law of thermodynamics: Heat does not flow spontaneously from a colder body to a hotter body; that is, work is required to cause heat to flow from a colder body to a hotter body. There are many equivalent statements of this law, each made by a different scientist at a different time. The law says that many processes in nature are irreversible, never going backward: burnt fuel is lost forever, an omelette cannot be turned back into eggs, isolated machines cannot stay in perpetual motion, and so on. It also defines the direction of time (time cannot go backward).

- Third law of thermodynamics: It's impossible to cool an object to a temperature of absolute zero ($-273.15°C$).

A perpetual motion machine is a machine that would run forever without consuming energy. It's impossible to build such a machine because perpetual motion violates the laws of thermodynamics. It defies the sacrosanct law of conservation of energy (the first law of thermodynamics), that no machine can produce more energy than it uses. The second law of thermodynamics places constraints on machines, such as car engines, that do useful work by tapping heat energy from burning fuel. This law demands that heat must flow from a hotter to a colder body. This means the machine must lose some energy when heat is converted into useful work.

42. Hydrocarbon

A hydrocarbon is a chemical compound containing only carbon and hydrogen. They are four types of hydrocarbons: (1) Saturated hydrocarbons (alkanes) contain only a single bond between the carbon atoms; they are called saturated because their molecules contain the maximum number of hydrogen atoms. (2) Unsaturated hydrocarbons (alkenes and alkynes) contain one or more double or triple bonds between the carbon atoms; alkenes have a double bond, alkynes a triple bond. (3) Cycloalkanes contain one or more carbon rings to which hydrogen atoms are attached. (4) Aromatic hydrocarbons (arenes) have alternating double and single bonds between carbon atoms forming rings.

Hydrocarbons can be gases, liquids or solid. They are present in all types of fossil fuels – petroleum, natural gas and coal – and therefore they are the primary source of energy.

Alkanes

Straight-chain and branched alkanes have the general formula C_nH_{2n+2}, where n is the number of atoms; for example, methane, CH_4, ethane C_2H_6.

Petroleum or crude oil is a mixture of alkanes which yield a wide range of combustible products after fractional distillation. Petroleum and natural gas (it may be found alone or with petroleum and is mainly mixture of methane, ethane, propane and butane – all alkanes) are fossil fuels because they began as organic life materials in the sediments of inland seas or coastal marine basins. The sources of this organic material are thought to be plankton (microscopic organisms in seas) as well as plant and animal debris swept from the land by rivers. As these carbon compounds sink deeper into the earth under accumulated sediments they are subjected to increasing temperatures and pressures and undergo chemical reactions. Oxygen and other elements are eliminated, resulting in a mixture composed almost entirely of hydrocarbons. These hydrocarbons migrated from source rocks to become trapped in large underground beds beneath layers of impermeable rocks.

Petroleum is never used as it is. It is separated into useful products by fractional distillation: gas, petrol, kerosene, heating oil, diesel, lubricants, paraffin wax and asphalt.

Alkenes and alkynes

Straight-chain and branched alkenes have the general formula, C_nH_{2n}. Common alkenes are ethene C_2H_4, propene C_3H_6, butene, C_4H_8 (all gases) and pentene, C_5H_{10} (liquid).

Straight-chain and branched alkynes have the general formula, C_nH_{2n-2}. The first and the most important member of the family is ethyne (acetylene), C_2H_2, a gas used in oxyacetylene burners for cutting and welding.

Cycloalkanes

Cycloalkanes are saturated hydrocarbons. Rather than being connected in a chain (like alkanes, alkenes and alkynes), cycloalkanes have one or more rings of carbon atoms. The general formula for cycloalkanes with one ring is C_nH_{2n}; for example, cyclopropane C_3H_6.

Aromatic hydrocarbons

All aromatic hydrocarbons contain one or more benzene rings. Benzene, C_6H_6, is the simplest aromatic hydrocarbon. The six carbon atoms, each with one hydrogen atom attached, are arranged in a form of a flat hexagon. The sweet smell of benzene gave rise to the name aromatic for all members of the class, though many are odourless. Naphthalene (moth balls), another common aromatic hydrocarbon, has two benzene rings.

43. Immune System

Immunity is the ability, especially in animals, to resist dangerous microorganisms that enter the body. Immunity may be innate (natural or nonspecific immunity) or acquired (adaptive or specific immunity). Natural immunity is due to the body's inherent ability to produce millions of antibodies, specialised protein molecules that specifically bind to their target molecules or antigens. In the acquired immunity the body acquires the ability to produce the specific antibody. This usually arises from previous exposure to infection by the specific antibody or deliberately by vaccination. The human body's immune system is a whole collection organs and tissues which fight infections.

How does the normal human body defend itself against the onslaught of microbes? When viruses, bacteria or other microbes enter the body they dominate and assume the life of some cells and began to multiply. The intruders are first opposed by scavenger cells, known as phagocytes, which simply engulf and digest the microbes. Phagocytes do not engulf microbes at random: if they did so they would damage the body's own cells. Special chemicals on bacterial wall cells stimulate phagocytes into action. Phagocytes make up the body's innate immune system, the simplest type of immune system. The innate or nonspecific immune reactions may be followed by acquired or specific immune reactions.

There are two main types of phagocytes. Polymorphonuclear neutrophils, or PMNs, make up 60 per cent of white blood cells. The bone marrow produces 80 million of PMNs every minute, but they survive only a few days.

Other scavenger cells, called macrophages, are long-lived. They make up only 6 per cent of the white blood cells, but combine the whole immune system to reject the invaders. After engulfing the invader, macrophages display specific markers, or antigens, from the invading microbe on its surface. T cells, members of a group of cells known as lymphocytes found in the bone marrow, recognise the antigen and grow into three types of populations.

The first type of T cells, helper cells, help B cells, which are also a type of lymphocytes, in producing antibodies. When the B cell encounters an organism bearing an antigen it has been programmed to recognise, it divides rapidly and produces plasma cells, which are effective hosts for synthesising its particular antibody. A single plasma cell can make 200 antibody molecules in one second, which bind to the antigen. This process alone destroys the antigen and, along with it, the invading organism.

Antibodies also activate the complement system, a group of proteins that occur in the blood and participate in immune reactions.

The second type of T cells, killer T cells, recognise and destroy virus-infected cells. The third type of T cells, suppression T cells, suppress the production of antibodies after infection has been overcome. B and T cells form memory cells, memory lymphocytes, that circulate in the body for years, sometimes for a lifetime. This becomes the first line of defence against future infections.

Specific immune responses can also be triggered by deliberately introducing a particular antigen in the body. The body mounts a counterattack and develops an army of memory cells, which destroy any antigen that may later invade the body. The more memory cell the body has, the more resistant the body is to a future infection.

This is the principle behind immunisation. The preparation used to provoke memory cells is known as a vaccine. A vaccine mimics the organisms that cause the disease, alerting the immune system that certain microbes are enemy agents. Because of this advance warning, the memory cells are ready to attack before the disease has time to develop.

Most vaccines in use can trigger immune responses in three ways. The first type of vaccines is made up of killed microorganisms. The microorganisms are killed (by heating or by placing them in formaldehyde) in such a way that the structure of the molecules on their surface is preserved. These microorganisms cannot replicate inside the recipient's body, but are capable of stimulating the production of antibodies. Whooping cough, cholera, influenza and (injected) Salk poliomyelitis vaccines are made of killed microorganisms.

The second type of vaccines contain genetically attenuated (altered or weakened) microorganisms that can still infect the recipient and stimulate immune response but do not generally produce disease. Smallpox, measles, tuberculosis and (oral) Sabin poliomyelitis vaccines contain mutant microorganisms.

In diseases such as tetanus and diphtheria it is not the microorganisms that kill but the toxins, poisonous chemicals, they release into the blood stream. The third type of vaccines acts against toxic products of microorganisms. They are made from inactivated microorganisms.

44. Infinity

Infinity refers to something that is endless, limitless and unbounded. In mathematics, infinity is a number – although the weirdest we know. Infinity is a mind-boggling concept: 'If you remove a part from infinity or add a part to infinity, still what remains is infinity,' according to an ancient Indian mathematical text. Mathematicians represent infinity with the sign, ∞, known as a lazy eight. They prefer to call it the lemniscates. The symbol was introduced in 1655 by the English mathematician John Wallis. He used it as shorthand, which is still used in calculus, for the phrase 'becoming large and positive'.

Though infinity fascinates mathematicians, the eminent British astrophysicist Arthur Eddington was no fan of it: 'That queer quantity "infinity" is the very mischief, and no rational physicist should have anything to do with it. Perhaps mathematicians represent it with a sign like a love-knot.'

The famous infinite hotel paradox devised by the famous German mathematician David Hilbert provides a peek into the weird world of infinity.

Hotel Infinity has an infinite number of rooms; all lined up in an endless corridor and numbered 1, 2, 3, ... forever. One night all its rooms are occupied, yet the 'Vacancy' sign is still on. A new guest arrives and asks for a room. 'No worries,' smiles the wily proprietor and he moves guest occupying Room 1 into Room 2, the occupant of Room 2 into Room 3, and so on. He now asks the new guest to move to Room 1, which becomes vacant when all his guests have moved to their new rooms.

Next night a tourist bus, with an infinite number of tourists on board, arrives at the hotel. 'No worries,' shouts the proprietor, 'just wait a minute.' With infinite patience he moves the guest in Room 1 into Room 2, the guest in Room 2 to Room 4, the guest into Room 3 into Room 6, and so on, leaving all the odd-numbered rooms vacant for the infinite number of newly arrived guests.

Hotel Infinity will not even run out of rooms for an infinite number of tourist buses, each packed with infinite number of tourists, if they all arrive at the same time. In the world of infinity, a part can be equal to the whole.

Here are some more strange properties of infinity (n is an ordinary number) to baffle you:

- $n + \infty = \infty$
- $n - \infty = \infty$
- $n \times \infty = \infty$ (if n is not equal to 0; if n is negative, $n \times \infty = -\infty$)
- $\infty/n = \infty$
- $n/\infty = 0$
- $n/0 = \infty$ (if n is not negative or equal to 0). For practical purposes it is not a 'legal' fraction.
- $\infty + \infty = \infty$
- $\infty \times \infty = \infty$
- $\infty - \infty$, $0 \times \infty$, ∞/∞ all give 'undefined answers. These operations are not allowed.

45. Intelligence

It is very hard to define intelligence. Dolphins (capable of abstract communication), primates (can use simple tools) and African Grey parrots (can categorise objects), for example, are intelligent in their own ways, but they all lack the most important aspect of human intelligence: its creativity that has resulted in the development of technology. If technology defines the intelligence of humans as a species, what defines the intelligence of individual humans? Nobel Laureate Eric R. Kandel, pinpoints it 'to be able to think deeply about problems, to be able to analyse new problems, to see relationships between events, to be creative.'

What does it really mean to be intelligent?

There's no absolute measure of these characteristics of intelligence. The most famous – not necessarily the perfect – measure are IQ tests. IQ or intelligence quotient, is the ratio of actual mental age, as measured by intelligence tests, to the mental age that is normal for a particular chronological age – 70 per cent of people have an IQ between 85 and 115.

As real-world intelligence has many dimensions, the tests fail to measure all of them. We can be more intelligent in some things and not in others. No one is equally intelligent in everything.

Intelligence is not related to the size of the brain or the numbers of neurons in the brain. There is no single intelligence centre in the brain; intelligence is built by a network of regions across both sides of the brain. Intelligence is how fast neurons are making connections, how well information is travelling throughout the brain. Richard Haier, an American psychologist who has been using brain imaging to discover the neural basis of intelligence, says intelligence is characterised by individual differences in learning, memory and attention and how they are integrated in any one individual, and the brain can generate the same IQ scores a number of ways. He believes that one day we may be able to estimate someone's

IQ and other intelligence factors from a brain scan.

Is there a difference between lower and higher IQ brains? Brain scans do show a difference. The lower IQ brains show lots of activity: they try harder. The higher IQ brains show less activity: they try smarter, not harder. 'It's easy for them, relatively speaking,' says American cognitive neuroscientist John Gabrieli. 'Smarter brains, simply put, are more efficient.'

No gender differences

You cannot tell male and female brains apart just by looking at them. There are subtle differences. Men's brains are slightly bigger than women's brains overall, even when the size is corrected for the fact men are on average taller than women. As brain size is no indicator of intelligence, men and women show no consistent difference on IQ tests. Some anatomical and physiological differences also occur in male and female brains. For example, brains of males produces neurotransmitter serotonin at a faster rate than those of females. Serotonin influences mood, so it may explain why women are diagnosed with depression twice often as men.

However, there is no convincing evidence that these differences result in differences in cognitive abilities. Whatever the differences are, they are biological; they are not innate or hardwired. Now we know about neuroplasticity and how experience can change the brain. 'Because gender traits are influenced more by a partiality for typically masculine or feminine interests, jobs and fashion styles than they are by biological sex, it follows that our malleable brains reflect these experiences,' says American neuroscientist Lise Eliot.

Though there are some behavioural differences between the sexes, they have been overblown, mostly by some educators. One of these differences is spatial reasoning: men show some advantage at mental rotation, which makes it easier to compare two-dimensional or three-dimensional shapes when rotated in space. Put simply, men are better at reading maps. But many wives will testify: they are better skilled than their husbands in reading maps. Some studies have reported that women are more likely to use landmarks and men more likely to uses distances and geometry when navigating. Differences in spatial ability have not been noted in young children, and studies show that spatial ability can be improved with training. There is no 'scientific proof' that says girls cannot successfully pursue careers in mathematics, physics and engineering.

Women also show an advantage in language on average (many studies and endless text-messaging by girls seem to prove it). But again, this difference hasn't stopped men becoming great writers. 'Kids rise and fall according to what we believe about them,' notes Eliot.

Beyond human intelligence

Artificial intelligence is the science of making machines that can reason like humans. 'If we can imagine an artefact that can collect, assemble, choose among, understand, perceive, and know, then we have an artificial intelligence.' That's how Joseph Feigenbaum, an American AI pioneer, defined AI in 1983. We're a long way away from seeing true AI.

Are we alone in the universe? Are there other civilisations beyond Earth? *Extraterrestrial intelligence* is an idea that fascinates many of us. Unlike UFOs, the search for extraterrestrial intelligence is a scientific endeavour.

46. Internet

The internet is a network of millions of computers around the world. The internet is not an individual organisation or network but a collective term for millions of networks connecting more than billions of users in almost all the countries. It has no central computer: each message bears an addresses code that lets any computer in the network forward it towards its destination. Machine that store the information are called servers. Every machine follows a set of rules to complete tasks. Without a common set of protocols machines connected to the internet would not be able to understand one other.

The most important protocol for the internet is TCP/IP (transmission control protocol/internet protocol). It describes how an internet-connected computer should break down data into packets (parts of a file between 1000 and 1500 bytes), and provide addresses to routers to move those packets across networks to their destination. Each device connected to the internet has an IP address, which is a 32-bit binary number and has two parts – a network identifier, and a host identifier. As we are running out of addresses a new system of 128-bit addresses has now been developed. IP addresses are typically given in four-part decimals numbers from 0 to 225 separated by dots, each part represents 8 bits of the 32-bit address; for example, 134.218.18.26 .

World Wide Web (WWW or simply the Web) is a hypertext-based graphical information and resource system on the internet. All document files on the Web are in hypertext and can be linked to other files on the same computer or to files on another computer. Hypertext is an information retrieval system presented in the form of a text file. It uses highlighted keywords, called triggers, embedded within the text which, when selected either pressing a particular key or clicking the mouse, take you to information on that subject. The information may contain other triggers which route to other subjects.

In the 1980s a British computer expert, Tim Berners-Lee, while working at

CERN, the joint European particle physics lab in Geneva, developed a simple programming language that he called HTML or HyperText Markup Language. HTML contains simple codes (such as 'this text has <B> **bold** </B>and <I> *italic* </I> words') that are used to format text and include graphics, audio and video. He also designed a protocol (HTTP, or HyperText Transfer Protocol) to move files across the internet, and a system of addresses (URLs, or Uniform Resource Locators) to locate a file on the internet. All he now needed was a way to view HTML files. He devised a simple browser program, which he called World Wide Web and on 6 August 1991 unveiled the results on the CERN computers. The first web page (http://info.cern.ch/hypertext/WWW/TheProject.html) had no graphics, no dynamic images, no sound, no videos, just plain text. The rest is history.

47. Life

Although it is difficult to define life, everyone agrees that all living things reproduce and for this they must have a system for storing and duplicating instructions about their structure and passing it on to their offspring. For life on Earth, DNA provides such a system. Its most remarkable feature is its genetic code, which is the same for all life forms on Earth. This feature makes the genetic code as old as life itself. All living things also (a) grow and develop; (b) take in energy; (c) get rid of waste; and (d) evolve in response to their environment.

Our planet was formed about 4.6 billion years ago from a ring of gas and dust around the young Sun. For nearly 700 million years the young Earth was subjected to intense bombardment by gigantic asteroids, the debris left over from the formation of the solar system. Life appeared about 3.8 billion years ago, as soon as the cosmic bombardment had ended. Some biologists say that life on young and inhospitable Earth appeared 'fully formed, with almost indecent haste'.

Fast or slow, how did life begin?

In 1906 Svante Arrhenius, the Swedish chemist who gave us the chemistry of ions, for which he won the 1903 Nobel Prize, suggested that life did not begin on Earth at all, but arrived from the outer space. He said that microorganisms from another planet where life already existed travelled through space on a meteorite that finally landed on Earth, where they began to grow and develop. He called this process *panspermia* (Greek for 'all seeds'). Some scientists, though small in numbers but highly vocal, still believe in the idea of life from outer space.

In 1924 Aleksander Oparin (1894–1980), a young Russian biochemist, presented the first viable theory of the origin of life by natural physical and chemical means here on Earth. He suggested that early in the Earth's history the atmosphere was rich in hydrogen. Simple inorganic hydrogen compounds such as water, methane and ammonia could form organic compounds. Gradually these organic compounds fell from the atmosphere to the ground, where the rain – which occurred when Earth cooled and water vapour condensed – washed them into pools and ultimately into oceans. Over millions of years the organic molecules in this 'primordial soup' joined together into long chains of proteins and DNA molecules until a cell appeared which possessed the right kind of reactions and right kind of compounds to be considered an organism. This first cell could replicate itself and therefore it filled the bill for the first living organism.

In 1953 Stanley Miller (1930–2007), then a young graduate student working in the laboratory of Nobel Prize-winning chemist Harold Urey at the University of Chicago, provided the first experimental support to the 'primordial soup' theory. He subjected a mixture of methane, ammonia, water vapour and hydrogen to a series of electrical charges. He imagined this to be a rough duplication of conditions on the primitive Earth when the primordial soup was subjected to bolts of lightning. After a week, the inorganic molecules had joined to form several amino acids, building blocks for proteins which make up cells. Urey was exultant: 'If God didn't do it this way, He missed a good bet.' Miller's 'primordial soup' has been the staple diet of biology textbooks for decades, but is now being challenged by other theories.

The story of life as we know it is the story of water and carbon as they are the essential ingredients of life on our planet. Without them there would be no life on our planet, but could there be life without water or carbon (or both) on other planets? That we do not know – yet.

48. Light

Light is visible electromagnetic radiations. They range from about 390 nanometres to about 750 nanometres in wavelength. In a vacuum light travels at a speed of 299,792 kilometres per second (or 299,792,458 metres per second). In 1905, in his theory of special relativity Einstein said that the speed of light in a vacuum is constant and is independent of the speed of observer. In other words, Einstein's speed rule prohibits anything possessing mass travelling faster than light. According to this rule and modern accepted theory, the speed of light is absolute, and represents a limit which no object can exceed.

Isaac Newton was the first to attempt a definition of the nature of light. In 1704 he postulated that light consists of tiny particles called corpuscles which travel in 'aether of space'. However, Dutch scientist Christian Huygens, a contemporary of Newton, said that light consists of waves having wavefronts perpendicular to the path of light rays. The controversy was not resolved until 1865 in which year Scottish physicist James Clerk Maxwell developed a series of equations from which he showed that radiant heat and light travels in free space as waves. In 1905 Einstein said that light is transmitted as tiny particles, or photons as they are now called, rather than as waves.

The current view of the nature of light based on quantum theory – a theory built on the concept by German scientist Max Planck of the discontinuity of energy. He suggested in 1990 that energy is conveyed in discrete packets, or quanta. From this, light, as an electromagnetic radiation, is transported in photons that are guided along their path by waves. This is called wave–particle duality. In 1932 French physicist Louis de Broglie extended the idea of wave–particle duality by saying that under some conditions particles such as electrons might also behave like waves.

The visible light spectrum

The visible colours from shortest to longest wavelength are: violet, blue, green, yellow, orange, and red. Here're the approximate range of wavelengths of the colours:

Colour	Wavelength in nanometres
violet	380–450
blue	450–495
green	495–570
yellow	570–590
orange	590–620
red	620–750

Indigo is no longer distinguished as a separate colour (The popular mnemonic VIBGYOR used to remember the colours of spectrum in the reverse order has lost its currency.)

White light is a mixture of all these colours. It's not necessary to provide all these colours to give the sensation of white light. Red, green and blue add up to make up white light. They are primary colours; they cannot be made by mixing other coloured lights. Black is a total absence of all colours (or wavelengths). Monochromatic light is light of narrow frequency.

49. Magnetism

Magnetism is the phenomenon associated with magnets and magnetic fields surrounding them. A freely suspended magnet will set itself pointing along a position close to the north–south direction. The end that points towards the north is called the North pole, and the other end the South pole – the basic principle of the compass. Like poles of two magnets repel each other, while unlike poles attract. The poles always exist in pairs. A magnetic field – a region of space in which a magnetic force acts – surrounds every magnet. A magnetic field also surrounds Earth, which extends 60,000 kilometres in space.

The Earth's magnetic field (also known as geomagnetism) probably comes from slowly circulating mass of molten iron and nickel in the outer core.

The Earth's magnetic poles are gradually moving. The Earth's magnetic axis is at present oriented at an angle of 11° to the planet's axis of rotation. This means north and south poles and the equator differ from their magnetic counterparts.

The angle between the lines of force and the Earth is called magnetic inclination, or dip. It varies at different latitudes; from 0° at the magnetic equator to 90° at the magnetic poles. The angle between the direction of magnetic north (or south) and the corresponding geographical pole is called magnetic declination. It also varies from place to place.

The strength of the magnetic field at the Earth's surface ranges from 25 to 65 microtesla, which is hundreds of times weaker than the field between the poles of a toy horseshoe magnet. The Earth's magnetic field is weakening: it has decreased about 7 per cent since 1845.

Palaeomagnetism

'Fossil magnetism' or past magnetic field that is preserved in rocks. When certain rock are formed, magnetic particles in them line up in the direction of the Earth's magnetic field at the time. When such rocks are finally consolidated, the magnetic field becomes fixed, or fossilised, in them. The study of these rocks can reveal the orientation of magnetic field in the past, and this has been used as a further proof of the theory of continental drift.

Magnetic behaviour

Different materials show different behaviour when placed in a magnetic field:

- Diamagnetic materials become magnetised in an opposite direction to the magnetic field. They are repelled by a magnet, although this opposite effect is very small. Copper, silver and gold show diamagnetism.
- Paramagnetic materials become magnetised with their magnetic axes (the line joining the north pole to the south pole) parallel to the magnetic field. They are weakly attracted to a strong magnet. Paramagnetic materials include magnesium, molybdenum, lithium and tantalum.
- Ferromagnetic materials exhibit strong increase in magnetic field. Unlike diamagnetic and paramagnetic substances, they do not lose their magnetic properties completely when taken out of the magnetic field. They retain a weak magnetic field, or residual magnetism. Iron, nickel and cobalt show ferromagnetism.

50. Matter

Matter is anything that occupies space and possesses mass. It can exist in three physical states: solid, liquid and gas. Both liquids and gases flow, hence they are also known as fluids. The gaseous state of substances which exist as liquids at room temperature and pressure is called vapour. Thus we speak of oxygen gas and water vapour. Solids, liquids and gases differ in shape, volume and compressibility. Solids have fixed shape, fixed volume and no compressibility. Liquids have fixed volume but no fixed shape; they have negligible compressibility. Gases have no fixed shape or volume; they have high compressibility.

The kinetic theory of matter provides a model for each of the three states of matter. The theory assumes that matter is composed of particles called molecules with definite and characteristic sizes. The molecules are in constant random motion and interact with each other through attractions and repulsions. The distance between the molecules are very large compared with the size of molecules themselves. The velocity of molecules varies as the temperature varies as shown in the following diagram.

Sometimes the term *phase* is used synonymously with the states if matter. Phase transition is described as transition between, solid, liquids and gases. The various phase transitions are described in following diagram.

In physics, hot, ionised gas found in flames, lightning, fluorescent lights, the Sun, and the explosion of hydrogen bomb is known as *plasma*. Plasma is also called the fourth state of matter. More than 99 per cent of the mass of the universe is in the form of plasma.

Molecules in three states of matter

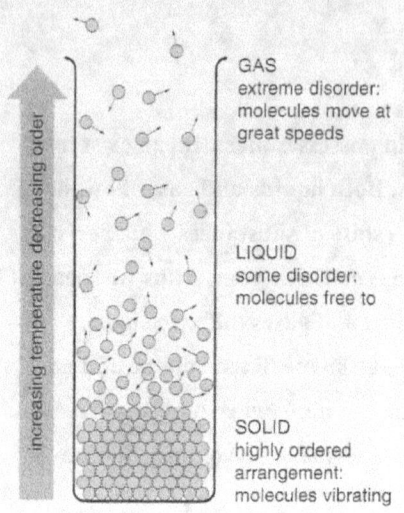

increasing temperature decreasing order

GAS
extreme disorder:
molecules move at
great speeds

LIQUID
some disorder:
molecules free to

SOLID
highly ordered
arrangement:
molecules vibrating

Various changes of states of matter

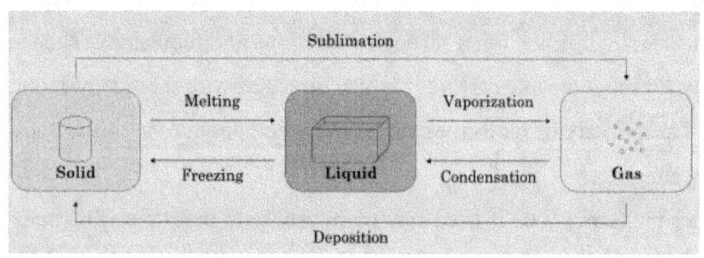

Sublimation

Melting

Vaporization

Solid

Freezing

Liquid

Condensation

Gas

Deposition

51. Mass

The mass of an object is the amount of matter it contains. It is a state of object's inertia, the tendency of an object to resist a change in its state of rest or motion. Mass is a measure of inertia. Weight is the force with which an object is attracted to the Earth. The weight of an object – whether it is falling freely or whether it is resting on a table – is equal to *mg*, where *m* is the mass of the object and *g* acceleration due to gravity. Mass is measured in kilogram (kg), weight in Newton (N).

In everyday usage, the words 'mass' and 'weight' are used interchangeably. In science, they are two different things.

The equation $F = ma$ relates mass to force (or weight). The world's most famous equation, $E = mc^2$, relates mass to energy. The equation shows that mass and energy are mutually convertible under certain circumstance.

For centuries scientists believed that energy and mass were two entirely separate things. In 1905 Einstein showed that mass and energy were one.

To appreciate the significance of $E = mc^2$, consider the following: E is energy in joules, m is mass in kilograms and c is the speed of light in metres per second (about 300,000,000). Thus, the energy released by 1 kilogram of matter

= 1 × 300,000,000 × 300,000,000 joules

= 90,000 million million joules

= energy released by 20,000 kilotons of TNT

The Hiroshima atomic bomb was only a 15-kiloton bomb. For years Einstein believed that energy could not be released on such a tremendous scale. The Hiroshima atomic bomb proved him wrong. The mere 15-kiloton blast of the Hiroshima atomic bomb on 6 August 1945 killed 140,000 people, injured hundreds of thousands and destroyed 70,000 buildings. It almost wiped the city of Hiroshima from the map of Japan.

52. Measurement

'A quantity like time or any other physical measurement does not exist in a completely abstract way,' writes British cosmologist Hermann Bondi in his book, *Relativity and Common Sense* (1964). We find no sense in talking about something unless we specify how we measure it ... a definition by the method of measuring a quantity is the one sure way of avoiding talking non-sense.' Metrology – the science of measurement – deals with processes that determine the ratio of a physical quantity such as length to a unit of measurement such as metre. Accuracy and precision are two important concepts in measurement.

In 1960 the General Conference on Weights and Measures gave the name International System of Units (usually known as SI from French *Système international d'unités*) to the metric system. The SI is now the universal system of measurement in science.

The SI is based on seven base units. Other units can be obtained from these base units by multiplication or division.

Physical quantity	SI unit	Unit symbol
Base units		
length	metre	m
mass	kilogram	kg
electric current	ampere[2]	A
thermodynamic temperature[1]	kelvin[2]	K
luminous intensity	candela	cd
amount of substance	mole	mol

1: By definition, one degree on the Celsius scale (°C) has the same size as one degree on the Kelvin scale

2: The unit symbol taken from a scientist's name is written in uppercase letter

All SI units, except kilogram, are now defined in terms of fundamental physical phenomena. For example, metre is derived from the speed of light, and second from the frequency of radiation emitted from the caesium 'atomic clock". Kilogram is only that depends upon an actual artefact. Kilogram is defined as the mass equal to that of the international prototype kept by the International Bureau of Weights and Measures at Sèvres, near Paris, France. The cylindrical prototype (*see accompanying photo*), known as *kilogramme des archives*, was made in 1889 and contains 90 per cent platinum and 10 per cent iridium. Its 39-milimeter diameter equals its height. Kept under tight security, the prototype is mysteriously losing weight. In 2011 the Bureau announced that the prototype appears to have lost 50 micrograms compared with the average of dozens of copies.

The table below shows numerical prefixes used in the SI.

Prefix	Symbol	10^n
yotta	Y	10^{24}
zetta	Z	10^{21}
exa	E	10^{18}
peta	P	10^{15}
tera	T	10^{12}
giga	G	10^9
mega	M	10^6
kilo	k	10^3
hecto	h	10^2
deca	da	10^1
deci	d	10^{-1}
centi	c	10^{-2}
milli	m	10^{-3}

micro	μ	10^{-6}
nano	n	10^{-9}
pico	p	10^{-12}
femto	f	10^{-15}
atto	a	10^{-18}
zepto	z	10^{-21}
yocto	y	10^{-24}

For a specified physical quantity, accuracy is the closeness of measurement to the true value. It is expressed as either as an absolute error (the difference between the measured value and the true value) or a relative error (the ratio of the absolute error and the measured value). Precision is the agreement among several measurements that have been made in the same way. A precise experiment has a small experimental error.

Numerical prefixes used in data storage

Byte is the basic unit for measuring the size of memory in computers. In each byte there are 2^8 (= 256) possible combinations of 0s and 1s. Each combination can represent a letter, a number or a space. In computer storage, SI prefixes are applied to base 2, not 10. For example:

1 kilobyte = 2^{10} or 1024 bytes
1 megabyte = 2^{20} or 1,048,576 bytes
1 gigabyte = 2^{30} or 1,073,741,824 bytes
1 terabyte = 2^{40} or 1,125,899,906,842,624 bytes

53. Metabolism

Metabolism refers to all of the chemical processes occurring with a living organism. These processes include both the synthesis, *catabolism*, and breakdown, *anabolism*, of compounds. In catabolism, large molecules such as carbohydrates, proteins and fats are broken down to yield simpler molecules such as amino acids and yield energy. In anabolism, the cells use energy to produce molecules it needs for growth and repair. Your body's metabolism controls the amount of energy your body burns at any given time. You cannot regulate your body's metabolism; the only control you have is on your body's energy expenditure, which increases with exercise.

Your body's *metabolic rate*, or total energy expenditure, has three components:

- *Basal metabolism rate* (BMR) is the minimum amount of energy needed to maintain life; or the amount of energy the body uses at rest to maintain function such as heart beat, breathing and temperature. BMR varies from organism to organism and with sex and age. In humans, basal metabolism rate accounts for about 50–80 per cent of total energy needs.
- Energy used during movement and physical activity (from lifting your arm to strenuous physical exercise). This energy accounts for about 20 per cent of total energy used by an average person.
- Thermal effect of food. Energy used for eating, digesting and metabolising food. This energy accounts for about 5 per cent of total energy used by an average person.

Put simply, your body weight is a result of catabolism (the amount of energy released from the food you eat) minus anabolism (the amount of energy your body uses up). The excess energy is stored as fat or glycogen. Glycogen is a carbohydrate consisting of chains of glucose units and is stored in muscles and liver cells. Gaining weight is mostly a result of the body storing excess energy as fat. However, the body's metabolism can be upset by genetic disorder, hormonal problems or other underlying medical conditions. If you gain weight, you cannot simply blame your metabolism.

54. Mind

Our minds (infinite and ethereal) seem very different from our brains (finite and material), but it is now generally accepted that mind and brain are really one. The great seventeenth-century French mathematician and philosopher René Descartes said that mind and body must be fundamentally different: the body was made of physical substance and occupied space whereas the mind was nonphysical and didn't occupy any space. His distinction between the physical and the mental – now known as Cartesian dualism – still has a major influence today. Many Eastern mystical traditions, however, teach that the mind and body belong to an individual continuum.

The human brain is gradually yielding its secrets to increasingly powerful tools that can image what happens inside a person's head, say, when doing mathematics or listening music. In recent years neuroscientists have scanned the brain with sophisticated big-name machines such as electroencephalography (EEG), magnetoencephalography (MEG), computerised axial tomography (CAT) and positron emission tomography (PET) and functional magnetic resonance imaging (fMRI) and have pinpointed numerous psychological functions to its specific parts.

Neurons are the basic working units of the brain. They transmit information to other neurons, muscle or gland cells, and they control body movement, perception, emotion and thought. The structure of the brain is complex, but here's a brief introduction to some of the important parts of the brain.

The limbic system in the midbrain controls automatic functions: for example, it tells you to pull your hand away from a flame. The prefrontal cortex right behind the forehead is known as the 'executive' region of the brain. It integrates information and allows us to makes decisions.

Four structures of the midbrain form what is known as the limbic system, which deals with urges and appetites. These structures are amygdala, hippocampus (important for forming new memories), thalamus (a kind of sensory relay station) and hypothalamus (regulates release of hormones). A network of neurons in these four structures – called the limbic loop – drives our decisions about whether or not to act on external and internal stimuli. A small pathway from the limbic system pumps the neurotransmitter dopamine into the frontal cortex, the 'executive' region of the brain right behind the forehead. When dopamine reaches the frontal cortex we feel good. Dopamine neurons play a vital role in brain networks that govern motivation and a sense of reward and pleasure; they are also associated with motor functions. These neurons go awry in Parkinson's disease, schizophrenia and drug addiction.

Undoubtedly, mental activity is a type of brain activity. Or, the mind is like software in the brain's hardware. But not everyone agrees with such mechanistic explanations of our mind. The mind–brain debate continues.

55. Mole

The mole – not of the underground or secret service variety – is the bane of chemistry student. Like metre, kilogram, second, ampere, kelvin and candela it is a base unit in the International System of Units (SI). It is used to measure a physical quantity called 'amount of substance'. But like metre, kilogram, second, ampere, kelvin and candela, the mole is not a counting unit. That's where the concept of the mole becomes confusing. In simple words, a mole is 6.02×10^{23} particles of anything. The number 6.02×10^{23} is known as Avogadro's number, a valuable constant in chemistry.

We can blame Latvian-born German physical chemist Friedrich Wilhelm Ostwald (1853–1932) for introducing the concept of the mole. In 1900 Ostwald, who was awarded the 1909 Nobel Prize for Chemistry, first used the term 'mole' (from the Latin *moles*, meaning 'a large mass') for the quantity of a substance whose mass in grams is numerically equal to its relative molecular mass.

To master the concept of the mole you must first understand the meaning of the terms 'relative atomic mass' and 'relative molecular mass'. The relative atomic mass (or atomic weight) is the average mass of one atom of an element, relative to the mass of a carbon-12 atom, taken as 12 exactly. The relative molecular mass (or molecular weight) of a compound is the sum of the relative atomic masses of all atoms present in the molecule of that compound. For example, the approximate relative molecular mass of H_2O is 18 ($= 2 \times 1 + 16$), where approximate relative atomic masses of H and O are 1 and 16 respectively.

Now suppose we have 1 gram of hydrogen. This amount of hydrogen will contain certain number of atoms, say n, each of which has approximate mass of 1. The relative atomic mass of oxygen is 16. This means that one atom of oxygen is 16 times heavier than one atom of hydrogen. Hence $n/16$ atoms of oxygen will be equal to 1 gram of hydrogen.

Or, the number of atoms in 1 gram of hydrogen equals n and the number of atoms in 1 gram of oxygen equals $n/16$. Or, the mass of n atoms of hydrogen equals 1 gram and the mass of n atoms of oxygen equals 16 grams.

The numbers 1 and 16 represent the atomic masses of hydrogen and oxygen respectively. Therefore, we can conclude that the atomic mass of any element expressed in grams will contain the same number of atoms. This number has been determined experimentally and is called Avogadro's number after the Italian physicist Amedeo Avogadro (1776–1856), noted for his work on gases. The value of Avogadro's number is 6.02×10^{23}.

Hence we can say that 1 gram of hydrogen contains 6.02×10^{23} hydrogen atoms; 16 gram of oxygen contains 6.02×10^{23} oxygen atoms; and so on. This concept can also be applied to molecules and ions.

Since 6.02×10^{23} atoms, molecules or ions is a very useful quantity for chemists, it is given a special name – the mole. A mole is 6.02×10^{23} particles of anything.

What about celebrating Mole Day from 6.02 AM to 6.02 PM on October 23 every year as many American high schools do. This day and time correspond to the size of one mole, that is, 6.02×10^{23}.

56. Molecule

A molecule is the smallest particle of a chemical element or compound which can exist by itself and retain the chemical properties of that element or compound. The molecule of a compound is formed when atoms of two or more elements combine together. The may be atoms of the same elements (for example, oxygen, O_2, and ozone molecules, O_3) or different elements (H_2O). An atom of an element may also be a molecule of the element (carbon atom and molecule, C). Molecules are held together by chemical bonds and are electrically neutral. They can vary greatly in size and complexity.

A molecular formula represents the actual composition of a compound. For example, the molecular formula for glucose is $C_6H_{12}O_6$. The subscript number to the right of the symbols tells the number of these atoms present in one molecule of glucose: 6 atoms of carbon, 12 atoms of hydrogen and 6 atoms of oxygen. A subscript of 1 is not shown: H_2O not H_2O_1. In the formula $Fe_2(SO_4)_3$, the sulphate ion, SO_4, is enclosed in parenthesis to indicate three sulphate ions are present, and the subscript 3 multiplies both the sulphur and oxygen atoms composing the ions (that is, 4 sulphur atoms and 12 oxygen atoms).

Some molecules contain tens of thousands of atoms. These giant molecules are called macromolecules. They are generally polymers which are formed by joining repeated smaller units of molecules. Smaller molecules are called monomers, the large molecules polymers. For example, glucose molecules (monomers) add on to each other to form a long chain which may contain as many as 300 glucose molecules. The result is cellulose, a natural polymer – the structure of plants. Polythene, a synthetic polymer, consists of more than 1000 monomer units of ethylene (or ethane), C_2H_4.

Molecules are always in constant random motion. Molecular motion approaches zero at temperatures approaching $-273.15°C$. This temperature is known as absolute zero. It is the theoretical lowest temperature. Like the speed of light, absolute zero can be approached closely but cannot actually be reached, as to reach it an infinite amount of energy is required.

At temperatures close to absolute zero atoms and molecules lose their separate identities and merge into a single 'super atom', a new form of matter. This super atom is known as Bose-Einstein condensate. It was predicted in 1924 by Einstein based on ideas originally suggested by Indian-born scientist Sateyndra Nath Bose. In 1995 American scientists were successful in creating Bose-Einstein condensate in a laboratory. Like solids, liquids, gases and plasma Bose-Einstein condensate is a state of matter.

57. Motion

Motion is a change in the position of an object with respect to time. The position, direction of motion and speed of an object describes its motion. Newton's laws of motion form the cornerstone of Newtonian or classical mechanics, which applies to the large-scale world; it does not apply to subatomic phenomena. In Newtonian mechanics, if the position and ordinary velocity of an object is known at a particular time, it can be calculated where the object will be at some point in future. But we cannot measure both the position and velocity of an elementary particle with absolute precision.

The three laws of motion stated by Isaac Newton are essential for understanding motion of bodies other than elementary particles. Newton presented his laws in his magnum opus *Philosophiae Naturalis Principia Mathematica* (Mathematical Principles of Natural Philosophy). The publication of *Principia* in 1687 is one of the most important events in the history of science. Generally considered the greatest scientific book ever written, it changed the world's view of the universe.

Newton's first law of motion

An object at rest will remain at rest and an object in motion will remain in motion at that velocity until an external force acts on the object. (In everyday language, speed and velocity are interchangeable terms having the same meaning. In physics, velocity has magnitude and direction. Speed, on the other hand, has magnitude only.)

This law introduces the concept of inertia, the tendency of an object to resist a change in its state of rest or motion. The inertia of an object is related to its mass.

Newton's second law of motion

The acceleration of a body is proportional to the external force acting on it and inversely proportional to the mass of the body.

This law explain the relationship between mass and acceleration, which can be stated by the equation, $F = ma$, where F is the force, m mass and a acceleration. Acceleration is the rate of change of velocity with time. The change can either be in magnitude of velocity (that is, speed) or direction of velocity, or both.

Newton's third law of motion

For every action there is an equal and opposite reaction.

This law shows that force always exists in pairs: whenever one object exerts a force on another, then the second object exerts an equal but opposite force on the first. The two forces are called action and reaction.

In quantum mechanics, things are different as it cannot predict specific events. It can only predict probabilities (the odds that something is going to happen or not going to happen).

58. Mutation

Mutation is alteration or damage to DNA molecule, which results in a permanent change in the chemical constitution of the chromosomes within an organism. Genes are segments of a DNA chain; they are units of heredity and are located in chromosomes. A changed or damaged DNA can change the genetic message carried by that gene. Mutations can occur naturally or they can be caused by external influences such as exposure to specific chemicals or radiation. Mutations can be beneficial, harmful or neutral. Natural mutations are random; they can result in beneficial variations within a population which can lead to evolution.

Cells multiply by division. When a cell divides it makes a copy of its DNA, and sometimes the copy is not identical. The small difference from the original DNA sequence is a mutation. This process occurs naturally.

Sometimes an external agent, called a mutagen, causes the DNA to break down. In such situations the cell tries to repair its DNA, but most likely it ends up with a repaired DNA that is not the exact copy of the original.

Here're some common mutagens:

Chemical mutagens

The first chemical mutagen, nitrogen mustard, was identified in 1942. Nitrogen mustard is a component of poisonous mustard gas used in the two world wars. Since then numerous chemical mutagens have been identified.

Ionising radiation

High-energy radiation from radioactive substance or X-rays removes electrons from atoms during its passage, leaving electrically charged particles (ions) in its path. When radiation hits living tissue, it not only produces ions but also free radicals that are more reactive than ions. Ionising radiation causes structural change within molecules such as DNA. A damaged DNA may replicate abnormally into an enormous number of identically damaged cells. Such cells are known as mutant cells. The mutant cells in turn replicate to form what is known as caner or malignant tumour. A carcinogen is an agent that causes cancer.

Ultraviolet light

Ultraviolet light, a component of sunlight, when absorbed by skin causes a crosslink to form between certain adjacent bases of the DNA. In normal cases the cells can repair the damaged DNA; but prolonged exposure to ultraviolet light can lead to mutant cells and then to skin cancer.

59. Nebula

A nebula (Latin for 'cloud') is a cloud of hot gas and dust between the stars. Some nebulae are regions where stars are being formed, while others are remnants of dead or dying stars (supernova remnants are also known as nebulae). There are four main types: (a) an emission nebula emits its own light; (b) A reflection nebula emits light from nearby stars; (c) A dark nebula blocks the space behind it; and (d) A planetary nebula is a gas cloud around a star at an advanced stage of evolution. The best-known nebula is the Orion Nebula in our galaxy.

The Orion Nebula is the central 'star' of Orion's sword (or the handle of the Saucepan). Another well-known nebula in our galaxy is the Crab Nebula. It is the remnant of a supernova observed by the Chinese in the year 1054. The pulsar in its centre rotates 30 times a second and its rotational energy is converted into the brightness of the nebula. The light from Crab Nebula, located 6500 light-years away in the constellation Taurus, reached Earth in 1054. It is now the best-studied object in the sky because of numerous observations made by many NASA satellites, including the Hubble Space Telescope.

A *supernova* is an old star that suddenly explodes as it blasts itself apart. The remaining matter forms a neutron star. A supernova explosion is believed to be the largest explosion in nature except for the formation of the universe itself. A supernova, now known as SN 1987A, was discovered in 1987 in the relatively near Large Magellanic Cloud, the closer and larger of the pair of galaxies visible to the naked eye in the far southern skies. SN 1987 was the first supernova bright enough to be visible to the naked eye since supernova SN 1604, which exploded in 1604 in our galaxy itself.

60. Neuron

A neuron is a nerve cell found in almost all animals. Neurons are the basic working units of the brain; they are the cells that actually create activity in the brain. They transmit information to other neurons, muscle or gland cells, and control body movement, perception, emotion and thought. In the adult human brain about one in ten of total cells are neurons. There are more than 100 billion neurons in the adult human brain, and each neuron is highly specialised and connects with up to 10,000 neighbours. Intelligence is not related to the numbers of neurons in the brain.

Neurons consist of a central cell body and a series of branches that extend into different parts of the brain to fire electrochemical signals from one another. There are two kinds of branches: axons conduct signals away from the cell nucleus; and dendrites receive coming information.

Neurons transmit electrical signals from one to another. These signals are carried by molecules across contact points, called the synapses, with other neurons. The molecules are called neurotransmitters; you probably have heard of some of them: dopamine, endorphin, histamine and serotonin. Brain activity is basically just a bunch of neurons firing. When one neuron fires up, it excites its neighbours, and they in turn fire up others giving rise to patterns of activity that result in thoughts, feelings and perceptions.

Not so long ago it was thought that the adult brain was an immutable organ in which damaged cells could not be repaired or replaced. The skin, blood, heart, lungs, kidney and lever all could do it, but not the brain. Neuroscientists have now found evidence that the brain can change even in adulthood, and sometimes these changes can take place within seconds. If certain neural pathways are blocked new pathways open up. As these new pathways continue to be used they become stronger and more prominent – a phenomenon captured in the oft repeated phrase 'neurons that fire together, wire together'. Neuroplasticity is the ability of our brains to produce new connections between neurons as a result of new experiences. In other words, neuroplasticity is how the brain changes, or rewires, with learning and experiences. The brain can also produce new neurons – not only in children but also in older people. This process is called neurogenesis.

Physical exercise can help the brain as it pumps more blood to the brain, which gives more oxygen to the neurons and thus making them better nourished. Exercise also causes the release of proteins known as growth factors, including one called brain-derived neurotrophic factor (BDNF). As the levels of BDNF build up neurons start to branch out and start building new connections in the hippocampus, a brain region important for memory. These new connections signify a new fact or skill which has been learned and stored for future use. As we age individual neurons start to die. We know now that this loss is not permanent: the brain can make new neurons. Again, BDFN plays a role in growing new neurons – and exercise helps in building up its levels by increasing blood volumes.

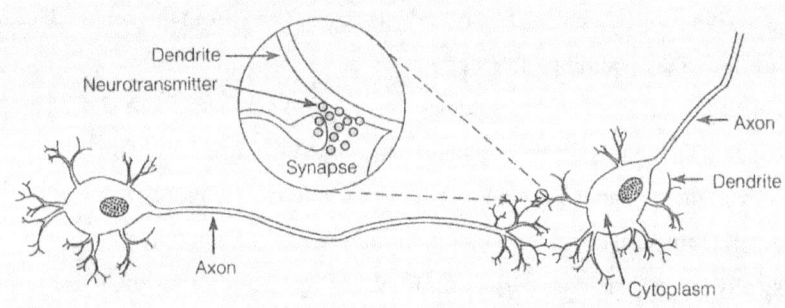

61. Organism

An organism is a living thing. Microorganisms are too small to be seen without a microscope (the words 'microorganism' and 'microbe' are often used interchangeably, but a microbe is a microorganism causing disease). All organisms are made up of cells. Their other features are: (1) They have specific size and shape. (2) They show some kind of movement which is directed and controlled for a certain purpose. (3) They are able to reproduce through sexual or asexual reproduction. (4) They show some kind of growth in their cellular mass or number of cells. (5) They can adapt to their environment.

In 1944 Erwin Schrödinger, the celebrated physicist famous for his 'cat' (not a real moggy, but a thought-experiment), wrote a little book, *What Is Life?*. Stepping outside his field of expertise, he speculated whether life is based on laws of physics. The book now available for free on the internet and still makes a fascinating read. A brief excerpt:

> The unfolding of events in the life cycle of an organism exhibits an admirable regularity and orderliness, unravelled by anything we meet with inanimate matter. We find it controlled by a supremely well-ordered group of atoms, which represent only a very small fraction of the sum total in every cell.

Classification of organisms

The cells of most organisms – *eukaryotes* – have nucleus surrounded by a membrane. However, the cells of some organisms – *prokaryotes* – do not have membrane-enclosed nucleus. Prokaryotes appeared some 3000 million years ago and then developed into more complex eukaryotes 700 million years ago.

Most taxonomists (scientists who name organisms and place them in groups based on their common characteristics) divide all organisms into five *kingdoms* (the first consists of all prokaryotes; the other four consist of eukaryotes.

- Monera – prokaryotes only (all bacteria and green-blue algae)
- Protista – unicellular eukaryotes such as slime moulds, algae and amoeba
- Fungi – simple structure, unicellular or multicellular eukaryotes such as mushrooms, moulds and yeast
- Plantae – complex structure, unicellular or multicellular eukaryotes such as mosses, ferns and flowering plants
- Animalia – complex structure, unicellular or multicellular eukaryotes such as humans, birds, worms, spiders and coral

Kingdoms are further divided into phylum (or division), class, order, family, genus, species. For example:

Classification of human	
Kingdom	Animalia
Phylum/Division	Chordata
Class	Mammalia
Order	Primates
Family	Hominidae
Genus	*Homo*
Species	*sapiens*
Scientific name: *Homo sapiens*	

Extremophiles

Extremophiles are the most fascinating organism or rather microorganisms. They thrive under conditions that would kill other creatures – in deep-sea hydrothermal vents, rock chimneys that grow above volcanic vents in the sea floor, through which erupts hot mineral-rich fluids; inside rocks buried kilometres below the Earth's surface where there is no oxygen, no organic food; in frozen Antarctic sea water; or in acidic, alkaline or saline environments.

Surface life is fed by photosynthesis (the process by which plants, algae and certain bacteria convert sunlight to chemical energy), but most extremophiles are fed directly by chemical energy. Both life forms have the same genetic system. This means that they have common origin.

Extremophiles offer the most probable model for testing the hypothesis that life exists on other planets. The reasoning is simple: if life can exist in extreme environments on Earth, it can also exist in the extreme environments of other worlds; for example, beneath the icy surface of Jupiter's moon Europa.

62. Osmosis

Osmosis is the diffusion of a solvent through a semipermeable membrane, which prevents any dissolved substances (solute) passing through it. A semipermeable is a thin, flexible partition – layer or film with tiny pores – which will allow the rapid passage of smaller solvent particles, but not of larger solute particles. Examples of semipermeable membranes in living things are: (a) the cell membrane, and (b) dialysis tubing. Osmotic pressure is the pressure that must be applied to a solution to prevent the flow of solvent, or osmosis, through the semipermeable membrane: it depends on the relative solute concentrations of the solutions involved.

For a living cell to survive concentrations of ions (particles with an electric charge) need to be the same on both sides of the cell membrane. As the cell membrane is selectively permeable, it allows water into and out of cells and organism. The movement of water across the cell membrane – osmosis – depends on osmotic pressure. Without the selectivity of the cell membrane, the substances essential for life of the cell would move out of the cell and toxic substances from the cell's surrounding would enter the cell. This would damage the function of the cell.

In a selectively permeable dialysis membrane, both the smaller solvent and solute molecules can pass through the membrane, but other larger molecules such as colloidal protein molecules are blocked. The process therefore separates proteins from small ions. The most important dialysis occurs in animal bladders.

Our kidneys clean blood by dialysis: they filter blood to remove harmful wastes and extra salt and water from the body. In case of kidney malfunction, some patients use a dialysis machine. In hemodialysis, blood from the patient is circulated through a long cellophane dialysis tube suspended in an electrolytic solution called dialysate, which contains the normal constituents of blood plasma. Toxic products such as urea from the blood diffuse through the dialysis tube, but other larger blood components remain behind.

Water and nutrients enter plants through the roots by osmosis. Osmotic pressure plays an important role in plants, which have a strong rigid cell on the outside of the cell membrane. The cell uses this wall to create osmotic pressure within it. The increase in pressure makes the cells rigid giving the plant a supportive structure. This is important, as plants do not have a skeleton. The pressure of water in the cells supports leaves and shoots.

Osmosis also explains how seawater and freshwater fish maintain water balance in their bodies. Seawater fish continually swallow water. This means their bodies have high salt and low water concentration. But the seawater outside has low salt- and high-water concentration. Therefore, water leaves the body of the fish by osmosis. In freshwater fish, water enters the body by osmosis. The reason is that their bodies have low water concentration because they produce large amounts of dilute urine. This makes water concentration high in freshwater outside.

63. Periodic Table

The periodic table is a table of chemical elements arranged in order of their increasing atomic numbers to show the similarities of elements with related electronic configuration. Atomic number is the number of protons in an atom; atoms with the same atomic number have the same chemical properties. In a neutral atom, the atomic number is always equal to the number of electrons. The atomic number of elements known so far range from 1 (hydrogen) to 118 (ununoctium – this is a temporary name until the discovery of element 118 is approved by the International Union of Pure and Applied Chemistry.)

The vertical columns of the table are called groups, and are numbered from 1 to 18. This is known as the standard or long form of the periodic table. In the older or the shorter form of the periodic table groups 3 to 12 are placed in one main group, making the number of groups eight, which are numbered in roman numerals, I, II, III, IV, V, VI, VII and VIII.

Some groups have characteristic names:

- Group 1: alkali metals (lithium, sodium, potassium, rubidium, caesium, francium)
- Group 2: alkaline earth metals (beryllium, magnesium, calcium, strontium, barium, radium)
  Group 17: halogens (fluorine, chlorine, bromine, iodine, astatine)
- Group 11: coinage metals (copper, silver, gold)
- Group 18: noble or inert gases (helium, neon, argon, krypton, xenon, radon)

The horizontal sequences or rows are called periods. They are numbered from 1 to 7. Period 6 contains a group of fourteen elements (elements 57 to 71) known as lanthanides or rare-earth elements because of their rare occurrence in nature. Period 7 has an apparently incomplete group of elements called actinides (elements 89 to 118). More elements are likely to be discovered in this group. The lanthanides and actinides are placed in two separate rows beneath the table.

The elements with atomic numbers more than 92 are called transuranium as they come after uranium (92). They are not found in nature and are synthesised.

Hydrogen – the lightest element, with atomic number 1 – has no obvious position in the periodic table.

64. pH

The pH describes how acidic or basic a substance is. The pH scale runs from 0 (most acidic) to 14 (most basic or alkaline). A solution is acidic when the pH is less than 7, and basic (alkaline) when the pH is greater than 7. The scale is logarithmic; for example, a solution with pH of 3 is ten times as acid as a solution with pH of 4. The pH (short for 'power of hydrogen') measures the concentration of hydrogen ions, H⁻, in water. Therefore, it can only be used for solutions of acids and bases in water.

The pH scale was devised in 1909 by Danish biochemist, Soren Sorensen. The pH expresses the negative logarithm (to the base 10) of the hydronium ion, H_3O^+ concentration.

$$pH = -\log_{10}[H_3O^+]$$

The pH of a solution may be determined by using a pH meter or indicators. A pH meter makes use of electrical conductivity of a solution to determine its pH.

Indicators are substances which change colour when acids or bases are added. Litmus, phenolphthalein, methyl orange and universal indicator (a mixture of several indicators) are the most commonly used indicators. The universal indicator changes colour gradually, from red through green to violet, as pH increases from 1 to 11. It may be placed in solution or on strips of paper which are known as pH papers.

The gastric juices in our stomach contain several hundred millilitres of dilute hydrochloric acid. This acid helps in the digestion of the food but too much of it can cause indigestion or heartburn. These symptoms usually arise when the pH of the gastric juices is between 1 and 3.

The acid in stomach can be neutralised with a base. Most health salts or antacids used for correcting indigestion or heartburn contain one or more of the following bases: sodium bicarbonate, magnesium hydroxide, magnesium hydroxide, aluminium hydroxide.

The pH Scale

increasing	0	battery acid
acidity	1	stomach acid
(6 to 0)	2	lemons, vinegar
Neutral	3	oranges, soft drinks
(7)	4	tomatoes, grapes
increasing	5	black coffee
alkalinity	6	most drinking water
(8 to 13)	7	pure water
	8	seawater
	9	baking soda
	10	milk of magnesia
	11	household ammonia
	12	washing soda
	13	oven cleaner
	13	sodium hydroxide

65. Photosynthesis

Photosynthesis (meaning 'combining by light') is the process by which plants and some species of algae trap and store solar energy. It converts solar energy into chemical energy needed by plants to change energy-poor compounds such as carbon dioxide and water into energy-rich carbohydrates and other organic compounds. Photosynthesis is nature's most important chemical process: it supplies oxygen to the atmosphere and food to plants which, directly or indirectly, supply energy to most living things. The process occurs in chloroplasts, specialised organelles found in plant cells. The chloroplast contains chlorophyll, a green pigment, which plays an important role in photosynthesis.

Two types of reactions take place in photosynthesis: light reactions in which photons (that is, solar energy) are absorbed by chlorophyll and converted into chemical energy, and water is split into oxygen and hydrogen. In dark reactions, carbon dioxide is converted into sugar.

The light reaction involves two phases which take place in two types of photosynthetic units. Each unit is made of several hundred light-absorbing pigment molecules of which chlorophyll is a principal component and one reaction centre. The pigment molecules act as antennae and harvests photons. A captured photon is transferred from molecule to molecule until it arrives at the reaction centre. This causes an excited reaction-centre molecule to release two electrons, which it replaces by tearing two others from a water molecule. As water molecule split, oxygen is released to the atmosphere but hydrogen remains in solution, which helps in the photosynthesis of hydrocarbons.

The overall chemical reaction is

$$6CO_2 + 6H_2O + (\text{solar energy}) = C_6H_{12}O_6 + 6O_2$$

Or, we can say that in the presence of sunlight six molecules of carbon dioxide react with six molecules of water to produce one molecule of glucose and six molecules of oxygen. Plants absorb water through their roots, and carbon dioxide through their leaves. Oxygen is released into the air as a by-product of photosynthesis. Some glucose is used for respiration (the process in which food molecules are broken down into simpler units with the release of energy), while some is converted into starch for storage.

Scientists are now looking at ways of creating artificial photosynthesis systems to harvest solar energy. Theoretically, solar energy can be tapped from these systems in two different ways: either as electrons (that is, electricity) or as hydrogen (a pollution-free fuel that can be used for heat or to make electricity). If scientists are successful in mimicking photosynthesis, we may one day have a have new source of renewable energy that does not pollute the environment

66. Planet

A planet is a celestial body that: (a) is in orbit around the Sun, (b) has sufficient mass for its self-gravity to pull itself into round (or nearly round) shape; and (c) has no other bodies in its path that it must sweep up as it goes around the Sun. Thus, the solar system consists of eight planets: Mercury, Venus, Earth, Mars, Jupiter, Saturn, Uranus and Neptune. Pluto does not qualify as a planet as it fails to meet criterion (c). Every 228 years, Pluto's unusual 248-years orbit brings it closer to the Sun than Neptune for about 20 years.

Dictionaries tell us that a planet is a large, round heavenly body that orbits a star and shines by the star's reflected light. This simple definition is good enough for you and me, but not for astronomers. A brief look at some of their ideas:

Planets are round: The first criterion for planethood should be roundness. If a body is big enough, its gravity will conquer geological and mechanical forces and will pull it into a spherical shape. The lower limit for diameter is about 800 km. Smaller objects – both asteroids and comets – are not round. Lawyers out there who want to quibble on the definition of roundness, should note that rotating planets are actually oblate; that is, they are flattened at the poles.

Planets orbit around stars in their own zones: The star is an object so massive that its interior burns with thermonuclear fusion. It may be a single star or a multiple-star system. A planet orbits around a star, but it orbits alone in its zone. This excludes asteroids, which swarm in an orbit between Mars and Jupiter. The criteria of roundness and orbit will add to the traditional roll-call of nine planets at least one asteroid – Ceres, the largest known asteroid.

Planets are less massive than thirteen times the mass of Jupiter: At about thirteen times the mass of Jupiter, the intense gravity of a body makes it hot and dense enough to start thermonuclear fusion. True stars are usually seventy or more times the mass of Jupiter. Any object more massive than thirteen Jupiters, would be a 'brown dwarf' star, and certainly not a planet.

In 2006 the International Astronomical Union ended the controversy about the definition of a planet once and for all by adopting three characteristics described in the shaded paragraph above.

For Pluto and similar objects, the IAU has created a new category called *dwarf planets*. This category also includes Ceres, Eris, Makemake and Haumea. There may be more than 100 dwarf planets awaiting discovery.

Dwarf planets beyond Neptune's orbit have been given a special name *plutoid*. Pluto and Eris are plutoids. They all reside in a belt of primordial icy bodies, known as the Kuiper belt, which surrounds the eight planets closest to the Sun. While all plutoids are dwarf planets, all dwarf planets are not plutoids.

The eight planets outwards from the Sun are:

- Mercury (with a furnace-like temperature of almost 500°C it's the hellhole of the solar system)
- Venus (seems like Earth's twin, but it's a parched and scorched world).
- Earth (liquid water, the elixir of carbon-based life, covers 75 per cent of its surface)
- Mars (it's no longer a scientific heresy to say 'life, not of the little-green-men variety, exists on the red planet today').
- Jupiter (one of its sixty-three known moons Europa fascinates astronomers more than the largest planet as it may have a deep ocean beneath its icy surface).
- Saturn (rainbows form after it rains on the ringed planet's largest moon Titan, but its 'water' is liquid methane; geysers of water spew out of reservoirs just metres beneath the icy surface of its tiny moon Enceladus hinting at the possibility of life).
- Uranus (a blue-green wonder with 11 separate rings and 27 known moons).
- Neptune (the smallest of the four 'gas giants', so called because they do not have solid surfaces; other three are Jupiter, Saturn and Uranus).

Then there are dwarf pants and also comets and asteroids, debris left over from the birth of the solar system.

67. Plate Tectonics

The theory of plate tectonics says that Earth's rigid outer shell, the lithosphere, consists of six major plates and several smaller ones (about 20 in all) which are in relative motion to each other. The plates are 70 to 150 kilometres thick and carry the continents and the oceanic basins on their backs like giant rafts. The plates are drifting like slow-moving ice floes over the mantle, the semi-fluid layer that underlies the lithosphere. Various plates are moving about 2 centimetres per year (which is about the same speed as our nails grow). Earthquakes are concentrated along the plate boundaries.

In 1915 Alfred Wegener, a German meteorologist, suggested that the continents had once been joined together in a giant supercontinent. He called this supercontinent Pangaea (Greek for 'whole earth'). It began to break away about 200 million years ago into the continents we know today, which slowly started drifting into their current positions. The mountains were formed when the edges of two drifting continents were crumpled and folded; and the oceans when they moved away from each other.

Ever since Darwin proposed that species are related by descent, scientists believed that 'land bridges' once connected the continents. The bridges, which allowed the species to cross the oceans, sunk when the planet cooled and shrunk. The idea of a supercontinent started developing in Wegener's mind when he noticed the jigsaw-like fit between the coastlines of Africa and South America. He presented his theory in 1915 in his book, *The Origin of Oceans and Continents*. The book became the most controversial, derided and ridiculed book in the history of geology. Part of the problem was that he never presented a convincing mechanism for his theory. Geologists discovered that mechanism in the 1960s and now Wegener's ideas are part of the theory of plate tectonics.

In 1962 the American geologist Harry Hammond Hess proposed his sea-floor spreading hypothesis: the floor of the ocean is continuously being pulled apart along a narrow crack centred on a 64,000-kilometre-long ridge – known as the mid-ocean ridge – that threads its way down the middle of the North and South Atlantic and across the Pacific and Indian oceans. Volcanic material rises from Earth's mantle to fill the crack and continuously create new oceanic crust.

Now the theory of plate tectonics (described in the box above) unifies the earlier ideas of continental drift and sea-floor spreading. The edges, or margins, of plates can move away from each other, push each other or slide past each other. Colliding plates form mountains. If they move away from each other oceans are formed. Sliding plates form mid-oceanic ridges. The San Andreas Fault in California is a classic example of sliding plates.

Earthquakes are the result of crustal rock breaking suddenly under the stresses caused by the continuous movement of continental plates. The point deep within Earth from which an earthquake shock wave comes is called the seismic focus. The point on the Earth's surface directly above the focus, where the intensity of shock waves is the greatest, is called epicentre. About 80 per cent of earthquakes occur around the perimeter of the Pacific. The bulk of the remainder occur along the Mid-Atlantic Ridge and along a belt that extends across the Mediterranean, the Middle East and India, connecting with the circum-Pacific belt from Japan to the Philippines. These earthquake areas correspond with the junction of continental plates.

68. Pollution

Pollution is addition of pollutants to the environment making it harmful to humans, animal and plant life. Pollutants are offensive, harmful and unnatural substances and cause short-term or long-term damage to the environment. The major forms of pollution are: air pollution, water pollution, plastic pollution, radioactive contamination, soil contamination, thermal pollution and noise pollution. Pollution affects us in many ways; for example, emission of carbon dioxide causes the greenhouse effect which is changing the climate, smog and haze affect human health and reduce the amount of light available to plants for photosynthesis, sulphur dioxide in air leads to acid rain.

Air pollution

The presence of damaging amounts of noxious substances in the air. Some air pollutants are gases; others are tiny particles (particulates). Fossil fuels are the major source of air pollution. Coal-fired power stations produce sulphur dioxide, oxides of nitrogen, carbon dioxide and particulates. Pollutants in the exhaust of a typical petrol-burning engine include carbon dioxide, carbon monoxide, methane, nitrogen oxides, sulphur dioxide, and particulates such as soot. Of these gases, nitrogen oxides are the main cause of photochemical smog, and sulphur dioxide causes acid rain.

Air pollution in some cities also depends on a climate effect known as *temperature inversion*. Normally, air temperature decreases with increase in height above the surface. Under some conditions a layer of cool air can be trapped under a layer of less dense warm layer. Pollutants in the trapped air cannot dissipate as long as the inversion lasts. Temperature inversion occurs more frequently in autumn and winter and last only a few hours.

Photochemical smog

A form of air pollution found in every major city where automobiles, or internal combustion engines, are wide use. This smog (usually see as combination of smoke and fog) is called photochemical because the chemical pollutants which cause it are produced in sunlight. Under the influence of sunlight, nitrogen oxides, which are present in the exhaust of a typical petrol-engine vehicle, convert oxygen into ozone. Ozone irritates human membranes and is harmful to vegetation.

Water pollution

Presence of pollutants in natural waters. Major types of pollutants are: microorganisms (source: sewage and animal waste); organic wastes (source: sewage, animal waste, decaying animals and plants, and discharge from food-processing factories); plant nutrients (source: chemical fertilisers); toxic heavy metals such as mercury, cadmium and lead (sources: industrial and chemical factories), sediments (source: erosion of soil by agriculture and strip mining); pesticides (source: chemicals used for eradicating insects, fungi and weeds); radioactive substances (source: mining of uranium-containing minerals); and heat (sources: cooling water used by power plants, which is discharged as hot water).

Acid rain

Rain containing enough acid to damage the environment. Rain is naturally slightly acidic – it reacts with carbon dioxide in the air to form carbonic acid with a pH of about 5.6. If the pH of the rain or snow is less than 5 – that is, more acidic than normal – it is called acid rain. The primary cause of acid rain is sulphuric acid, created by the reaction of sulphur dioxide with oxygen and water in the air. Most of the atmospheric sulphur dioxide is produced by burning coal with a high sulphur content. Acid rain increases the acidity of lakes and streams, damaging aquatic life. It also damages trees.

Plastic pollution

Plastic litter cannot be disposed of by burning (because it gives rise to too many harmful gases) or by landfill (because of its large volume). Recycling plastic is the best options for reducing plastic pollution. Other options are using photodegradable plastic (it breaks down when exposed to sunlight) or biodegradable plastic (it can be broken down by microorganisms) are. However, long-term environmental effects of degradable plastics are not yet fully known.

Noise pollution

Unwanted sound in our surroundings. Typical environmental noise may range from 40 decibels of a quiet office to the 140 decibels experienced at a distance of 140 metres when a jet aircraft takes off. A noise limit of 90 decibels at a distance of 1 metre is acceptable if the exposure to noise is limited to 2 hours.

69. Pressure

Pressure is force per unit area; that is, pressure = force ÷ area. The SI unit of pressure is pascal (symbol Pa), which is equivalent to a force of one newton per square metres (N/m^2). Atmospheric pressure is the pressure exerted by the Earth's atmosphere. It is defined as the force per unit area exerted against a surface by the weight of the air above that surface. It varies around the globe with temperature and weather conditions, decreases with increasing altitude. Atmospheric pressure at sea level is 101,325 pascal (101.325 kPa), which equals 760 millimetres of mercury or 1013.25 millibars.

Atmospheric pressure

For an observer in the northern hemisphere standing with his or her back to the wind, the atmospheric pressure will be lower to his or her left than to his or her right. In the southern hemisphere, the lower atmospheric pressure will be on the observer's right. In the northern hemisphere, the lower atmospheric pressure will be on the observer's right. This law – known as Buys Ballot's law – describes the relationship of the horizontal wind direction to the pressure distribution and can be used to work out the wind direction at different locations on a weather chart. The effect is caused by the deflection, caused by the Earth's rotation, in the movement of air from areas of high pressure to areas of lower pressure.

Pressure systems in the atmosphere

A *cyclone* is a region of low pressure in the atmosphere and is also called a depression. Lows are associated with fronts (a front is a boundary between two air masses) and storms. In the northern hemisphere wind circulation around lows is anticlockwise; in the southern hemisphere it is clockwise. A cyclone is indicated on a weather map by a LOW.

An *anticyclone* is a region of high atmospheric pressure which generally produces fine weather. In the northern hemisphere wind circulation around highs is clockwise; in the southern hemisphere it is anticlockwise. An anticyclone is indicated on a weather map by a HIGH.

A *tropical cyclone* is a spiral-shaped intense low-pressure system with extremely strong winds of up to about 200 kilometers per hour. It can range in diameter from 100 to 1500 kilometers. At the centre of the cyclone is a relatively calm region, called the eye. The more or less circular ring of high winds surrounding the eye is called the eye wall. Tropical cyclones form over all tropical oceans except the South Atlantic where a deep layer of moist air collects over warm oceans in the summer. A tropical cyclone is called a cyclone in the Indian Ocean and in the areas around Australia, *hurricane* in the eastern and central Pacific, the Atlantic Ocean, the Caribbean and the Gulf of Mexico and *typhoon* in the northwestern Pacific.

A *tornado* is a violently spiraling, small-scale column of air in contact with the ground. It typically forms from a cumulonimbus or thunderstorm cloud. It lasts only a few minutes but can reach speeds of 480 kilometers per hour.

Blood pressure

Hypertension, or high blood pressure, is called a 'silent killer' because people cannot sense or realise its symptoms. The chances of hypertension in a person rise with age, becoming more common after 35.

Blood pressure is the pressure exerted by blood in arteries. It is measured in millimetres of mercury (mm Hg). Blood pressure is always expressed in two numbers that represent systolic and diastolic pressures. Systolic is the pressure when the heart muscles contract and blood flows into the arteries. Diastolic is the when the heart muscles relax and the heart fills with blood from the veins. The blood pressure readings are always written one above the other, with the systolic number on the top and the diastolic number on the bottom. For example, 120/80 (120 over 80) mm Hg. Normal blood pressures (in mm Hg) for adults are shown below.

Category	Systolic	Diastolic
Optimal	<120	<80

Normal	<130	<85
High Normal	130-139	85-90
Hypertension		
Stage 1	140-159	90-99
Stage 2	160-179	100-109
Stage 3	>180	>110

70. Probability

Probability is the mathematical concept that deals with the chances of an event happening. Chance is something that happens in an unpredictable way. The probability of an event occurring is calculated by the number of possible occurrences divided by total number of likely outcomes. For example, when a coin is tossed, the likelihood of its coming down heads (or tails) is 1/2. The probability of an absolutely certain event is 1; the probability of an impossible event is 0. Probability can help you understand everything from your chances of winning a lottery to your chances of being struck by lightning.

The Chevalier de Méré, a seventeenth-century high-living French nobleman and gambler, liked to bet that a six would come up at least once when a die was rolled four times. But when he started betting that a six would come up at least once when two dice were rolled 24 times, he started losing money.

He asked his mathematician friend, Blaisé Pascal, why he was having bad luck in his new game. Pascal wrote to fellow mathematician, Pierre de Fermat, about this problem and their correspondence on the matter led to the birth of probability theory.

Below are some of the words we use to describe the likelihood of a particular event happening.

Description	Percentage	Ratio
Absolutely certain	100	1 in 1
Very likely	90	9 in 10
Quite likely	70	7 in 10
Evens (equally likely)	50	1 in 2
Not likely	30	3 in 10

Not very likely	20	1 in 5
Never (absolutely no chance)	0	0

As mentioned earlier, we can find the probability of an event by simply dividing the number of ways the event can happen by the total number of possible outcomes. This rule can be applied to tossing coins, rolling dice, dealing cards or drawing lottery numbers.

Take an example: When flipping a coin, what is the probability that it lands heads up when it hits the floor? There is only one way for the coin to land heads, so the number of choices is 1. The total number of options is 2 as the coin can either land heads or tails. So, the chance of a coin coming up heads is 1/2 or 1 in 2, which is the same as saying that the coin lands heads 50 per cent of the time. The probability of landing tails is also 1 in 2.

Take another example: What is the probability of drawing an ace of heart from a pack of well-shuffled pack of cards? There are four aces in a pack of 52 playing cards. The probability of drawing an ace is 4/52 or 1/13. The probability drawing an ace of heart is 1/52.

Combining probabilities

A die has six faces, numbered 1, 2, 3, 4, 5 and 6. The probability any one of these numbers comes is 1/6.

What would be the probability of getting either a 3 or a 5? Because 3 and 5 cannot occur together, such an event is called a *mutually exclusive event*. In mutually exclusive events, probability is calculated by adding individual probabilities. Therefore, the probability of getting either a 3 or a 5 is: 1/6 + 1/6 = 1/3.

When two dice are rolled separately, the second die does not take into account what the first die has done in order to decide what it will do. Such an event is called an *independent event*. In independent events, probability is calculated by multiplying independent probabilities. Therefore, when two dice are rolled separately, the probability of getting a double 6 is: 1/6 × 1/6 = 1/36.

Now, let's look at Chevalier's problem.

Single die

Probability of a 6 = 1/6

Probability of a number other than 6 = 5/6

Probability of no 6 in four rolls = 5/6 × 5/6 × 5/6 × 5/6 = 625/1296 = 0.48.

Probability of at least one 6 in four rolls = 1 – 0.48 = 0.52

Chevalier's chances of winning his bet were 52 per cent. The odds, as the gamblers say, were in his favour.

Two dice

Probability of a double 6 = 1/36

Probability of no double 6s = 35/36

Probability of no double 6s in 24 rolls = $(35/36)^{24}$

Probability of at least one double 6 in 24 rolls = $1 – (35/36)^{24} = 0.49$

Chevalier's chances of winning his bet were 49 per cent. The odds were against him.

71. Protein

Proteins are complex organic compounds that are present in all living cells. They are giant molecules made up by linking together a number of small molecules called amino acids (only 20 amino acids are found in proteins). All protein molecules contain carbon, hydrogen, oxygen and nitrogen. Some also contain phosphorus and sulphur. Plants can make amino acids and consequently proteins from inorganic compounds such as nitrates, water and carbon dioxide. Animals cannot make their amino acids. They depend upon plants or other animals for their supply. Common proteins include albumin (in egg white), keratin (in hair) and haemoglobin (in blood).

Proteins are building blocks of life. They are essential for growth and repair of all body tissue, and regulation of body processes. Proteins are digested to release amino acids. In the body, these amino acids are used to make new proteins, converted into hormones such as adrenalin or may be used as energy source. Animal proteins and vegetable proteins have the same effect on health.

Some sources of dietary proteins are:

- Meat, fish and poultry
- Soybeans, tofu and other soy-based products
- Eggs
- Milk and milk products
- Cereals
- Beans and lentils
- Seeds and nuts
- Whole grains, especially wheat

The human body can produce only 10 of the 20 amino acids present in cells. The others must be supplied in the food. Unlike fat and starch, the body does not store excess amino acids for later use; they must come from food every day. Failure to obtain amino acids would result in loss of body's proteins. Eating a variety of foods ensure that we get all the amino acids we need.

72. Pulsar

Pulsars (short for pulsating stars) are rotating neutrons stars that release regular bursts of electromagnetic waves. Pulsars are located by radio telescopes and found inside the debris of what is known as supernova remnant – leftover from supernova explosions. The discovery of pulsars – the first pulsar was discovered in 1967 – has enabled astronomers to study neutron stars, object never observed before. Neutron stars are only a few kilometres across and contain tightly packed neutrons, making them extremely dense. They do not glow and are so heavy that even a pinhead of their matter would have a mass of millions of tonnes.

In 1967 Jocelyn Bell (Burnell), a Cambridge University research student, had the sole responsibility for operating a radio telescope and analysing the data. The telescope – an array of radio detectors spread over an area of more than four acres – for the study of quasars (star-like objects that emit powerful radio waves). Quasar (*quasi*-stell*ar* object) is a compact and especially bright region in the centre of a massive galaxy surrounding its central supermassive black hole. Quasars are the most distant objects known in the universe and have been located by radio astronomers because of their powerful radio frequencies. A quasar is only the size of our solar system. Yet they are the brightest objects in the universe, the brightest emitting more light than 1000 galaxies or 10 billion stars.

After the first few weeks, Bell noticed some unusual markings – she called them 'scruff' – on the charts spewed by the telescope. Bell at once realised that they were definitely not from quasars. Closer examination showed that they were a series of intense pulses. Initially, Bell and Hewish thought the signals were from some extraterrestrial intelligence and dubbed them LGM (little green men). In fact, Bell had discovered the first evidence of a pulsar. There are now more than 1500 known pulsars.

73. Quantum

Quantum (plural *quanta*) is a quantity of something, a fixed amount. Energy is not a continuous quantity but it is quantised; that is, it could only take on certain values. When particles emit energy they do so only in discrete packets or quanta. For example, a photon is a quantum of light. This concept forms the basis of quantum theory. In popular usage, the term 'quantum leap' means a big jump. In physics, a quantum leap or quantum jump is a very tiny leap: it's an abrupt transition from one energy state to another in an atomic or molecular system.

Quantum theory, devised by German physicist Max Planck in 1900, led to quantum mechanics, the study of the mechanics of elementary particles.

Quantum mechanics does not replace Newtonian mechanics, it includes it. In Newtonian mechanics, if the position and ordinary velocity of an object is known at a particular time, it can be calculated where the object will be at some point in future. But we cannot measure both the position and the very high velocity of an elementary particle with absolute precision and therefore cannot specific events. The uncertainty principle, stated in 1927 by German physicist Werner Heisenberg, explains the reason why. To measure both the position and the momentum (which is, mass × velocity) of a particle (such as an electron) requires two measurements: the act of performing the first measurement will 'fix' a particle and so create uncertainty in the second measurement. Thus the more accurately a position is known, the less accurately can the momentum be determined. The disturbance is so small that it can be ignored in the macroscopic (large-scale) world but quite dramatic in the microscopic sphere. Quantum mechanics, therefore, can only predict probabilities (the odds that something is going to happen or not going to happen.

One of the bizarre features of quantum mechanics is quantum entanglement. All elementary particles such as electrons and photons vibrate. Consider the case of two electrons. Placed together they vibrate in unison. Place them apart, as far as another galaxy, and if you vibrate one of them, the other will immediately know the nature of its partner's vibrations and would dance to the same tune. Somehow the information between the two electrons is being transferred. Einstein called it 'spooky action at a distance' because the transfer of information could only be explained by assuming that it was travelling faster than light.

Could quantum entanglement be used to transfer information faster than light? The laws of quantum mechanics prohibit it, but it could be used to transfer information from one particle to another particle at a speed slower than that of light. In quantum teleportation, only the quantum state is teleported, not exact particles. So, quantum teleportation is not really sending instantaneously one particle, say an electron, from one place A to another place B. But the quantum states of the electron at A and B are indistinguishable. It's not really 'quantum faxing'; in faxing it's easy to tell the difference between the original and the copy. In quantum teleportation there is no difference between the original and the copy.

The most important application of quantum teleportation is in the field of quantum computing. Ordinary computing is based on the notion of bits. An ordinary bit can store only one number (0 or 1) at a time. In quantum state a particle can occupy two states at the same time, so a quantum bit, or qubit, can store two numbers at the same time. Thus each qubit doubles the size of a computer.

74. Quark

A quark is an elementary particle. Quarks are building blocks of matter; neutrons and protons are made up of quarks. Like electrons, they cannot be subdivided further. However, some physicists believe that quarks themselves are made up of even smaller particles. There are six types – known as flavours – of quarks: up, down, charm, strange, top and bottom. The proton is made up of up-up-down quark triplet; the neutron up-down-down triplet. Quarks cannot exist singly; however, they can be created in particle accelerators. All quarks, except the top quark, were created in 1977; and the tp quark was created in 1995.

Quarks were discovered by American physicist Murray Gell-Mann in 1964. When he was trying to coin a word for the new particle, he found the word 'quark' in Irish author's James Joyce's novel *Finnigans Wake* (1939):

Three quarks for Muster Mark!
Sure he has not got much of a bark
And sure any he has it's all beside the mark.

'Three quarks for Muster Mark!' implies 'three quarts of ale for Mister Mark'.

Gell-Mann predicted the existence of three quarks: up, down and strange. Another three were predicted by other scientists.

Quarks have the remarkable property of being permanently trapped inside particles such as the neutrons and the proton. These observable particles always have a whole number charge (such as 0, 1, –1 or 2, while quarks have a fractional charge, 1/3 or 2/3). Fractionally charged particles can only exist inside other particles

75. Radioactivity

Radioactivity is the spontaneous emission of extremely powerful radiations from matter without energy being supplied. This is the result of unstable atoms changing into stable atoms. Some radioactivity – natural radioactivity – happens constantly around us. Small amounts of radioactive atoms are found in the soil we stand, the food we eat, the water we drink and air we breathe. This radiation is known as 'background' radiation. There are significant differences in background radiation doses across the world. Our lifetime dose of background radiation is extremely low. Any dose of radiation, no mtter how small, involves a possible risk to human health.

There are three ways by which a radioactive atom can become stable. Each result in a different type of radiation – alpha particles, beta particles or gamma rays. A particular atom cannot emit both alpha and beta particles. If it emits alpha particles the process is called alpha decay, but if it emits beta particles the process is beta decay.

When the nucleus of an atom emits an alpha particle, it loses two protons and two neutrons and converts into the nucleus of a different element two steps down in the periodic table of chemical elements. An alpha particle is a helium nucleus, a tightly bound unit of two protons and two neutrons. For example, the alpha decay of thorium (element 90) produces nuclei of radium (element 88) and helium (element 2).

In beta decay one neutron changes into a proton, a neutrino and energy. Thus, the number of protons increases by one, while the number of neutrons decreases by one – the new atom moves one step up in the periodic table. This type of beta decay occurs in unstable nuclei because they have too many neutrons. For example, in the beta decay of radioactive carbon (element 6; ordinary carbon has 12 neutrons while radioactive carbon has 14) produces nitrogen (element 7), a neutrino and energy.

Two other types of beta decay occur when a material is bombarded with alpha particles:

- A proton can change into a neutron, a neutrino and a positron. This type of beta decay is known as positron emission.
- The nucleus captures an orbiting electron and the captured electron converts a nuclear proton into a neutron and neutrino. This process is called electron capture.

In gamma decay the number of protons and neutrons do not change the nucleus decreases its energy by emitting gamma rays, electromagnetic radiation of very short wavelength.

Gamma rays are highly penetrating and can only be stopped by tens of centimetres of lead or metres of concrete. They can cause serious and permanent damage to living tissues. Alpha particles can be stopped by a sheet of paper of human skin. Alpha particle sources are, however, a danger if inhaled or ingested. Beta particles are able to penetrate human skin but can be stopped by a thin piece of wood or plastic.

76. Relativity

Imagine a train speeding along a platform. According to classical physics, the length of the train would be the same for both a passenger on the train and a person on the platform. But according to special relativity theory, the train appears shorter to the person on the platform. This is because of a change in the nature of space cause by motion. Similarly, the time elapse of one second shown by two successive clock ticks will be one second to the passenger whether the train is stationery of moving, but to the person the platform it will be greater.

Moving clocks are measured to run slowly. Our measurements of time are affected by our motion. The rate at which the clocks run depends upon their relative motion. A person running away from a clock would observe it to move more slowly than his or her own clock. This is known as time dilation.

Now imagine the passenger on the train is holding a metre ruler. If he or she is pointing it in the direction of the train's motion, then the length of the ruler seen by the person on the platform will be less than a metre. The length of an object is measured to be shorter when it is moving than when it is at rest. The contraction in length is in the direction of motion and noticeable only when the speed of the object is a substantial fraction of the speed of light. For example, a metre ruler moving past us at a speed of 240,000 kilometres per second (that is, 80 per cent of the speed of light, which is about 300,000 kilometres per second) will seem only 60 centimetres long to us.

The theory also says that the mass of a moving object increases as it its speed increases. The increase in mass is noticeable only when the speed of an object is a substantial fraction of the speed of light. At the speed of light, the mass becomes infinite and therefore nothing can travel faster than light.

Time dilation, length contraction and relativistic mass are three outcomes of Einstein's theory of special relativity. Einstein found that his theory of special relativity is incompatible with Newton's laws of gravitation. To unite gravity with special relativity, in 1915 Einstein came up with his theory of general relativity. The theory recognises the impossibility of determining absolute motion and develops the idea of four-dimensional space-time continuum.

In brief, the theory of general relativity says that bodies do not attract each other by exerting a pull, but that the presence of matter in space caused space to curve in such a manner that gravitational field is set up. Gravity is the property of space itself. The theory also predicts that light should be bent by gravitational fields and time should appear to run slower near a massive body like Earth.

77. Reproduction

Reproduction is the biological process by which organisms generate offspring. Reproduction can be either sexual or asexual. Sexual production involves the joining of gametes to form a zygote. A gamete is a reproductive cell (sperm or egg, for example). When a male gamete (sperm) fuses with a female gamete (egg) the result is a single cell known as zygote, which develops into an embryo and finally into a new individual. In asexual reproduction, new individuals are formed from a single parent without gamete production. The offspring from asexual reproduction are genetically identical; the offspring of sexual production are genetically unique.

As in asexual reproduction offspring are genetically identical to the parent, there is less genetic diversity, which gives offspring a lesser chance of survival if the environment changes. In sexual reproduction, offspring have genetic characteristics from both parents. Thus, there is genetic diversity in sexual reproduction, which gives offspring a better chance of survival in changing environments.

Sexual reproduction is the primary method of reproduction for almost all macroscopic animals and plants. Single-celled organisms (such as yeast, bacteria, sea stars and sea anemones) reproduce asexually. Their offspring are formed by mitosis.

Inheritance of sex in humans

In mammals, sex chromosomes are a special pair of chromosomes, which is associated with the sex of the animal. Cells of women have two X chromosomes whereas those of men have one X and Y chromosome. An infant's sex is fixed by a single gene called *testes determining factor*, or TDF, which is in the male-determining Y chromosome. To create a male child, a father's sperm must carry a Y chromosome to fertilise a mother's egg, which always has an X chromosome. In the absence of Y chromosome – which means the absence of TDF gene – the embryo develops in a female child. (*Autosomes* is a general for all the chromosomes except the sex chromosomes. In humans, there are 22-pairs of autosomes – 23 pairs of chromosomes in all.)

Mitochondrial DNA

In animal cells, DNA resides in the nucleus where it controls the machinery of the cell. Like the nucleus, mitochondria – tiny organelles of the cell known as energy factories because they convert glucose into readily usable energy – also contain DNA. The nuclear DNA in human cells is in the form of chromosomes that are responsible for the transmission of genetic information. Because chromosomes come from both parents, nuclear DNA is shuffled and mixed at random with each generation, confusing the line of inheritance. But mitochondrial DNA is passed to the next generation only in the mother's egg cell – with no contribution from father because sperm's mitochondria do not survive fertilisation. Thus the mitochondrial DNA of a person is inherited from mother, maternal grandmother and so on. This feature makes mitochondrial DNA a very effective tool for relating individuals to one another. It also makes mitochondrial DNA easier to track though generations and to use it as a molecular clock. In 1987 a worldwide study of human mitochondrial DNA pointed out that all mitochondrial DNA stem from a woman who probably lived around 200,000 years ago in Africa. This so-called 'Mitochondrial Eve' hypothesis is still open to debate.

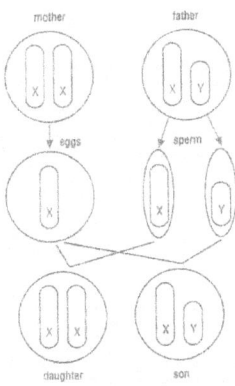

78. Respiration

Respiration is the process in plants and animals in which complex food molecules are broken down into simpler units with the release of energy. External respiration is the exchange of oxygen and carbon dioxide between the body tissue and the environment. Cell respiration refers to reactions of respiration in a cell; it occurs in both eukaryotic and prokaryotic cells. There are two basic forms of cell respiration. Aerobic respiration requires oxygen, which is provided by simple diffusion or breathing. It occurs in all plants and animals except yeast and bacteria. They rely on anaerobic respiration, which does not require oxygen.

happens in two stages:

1. Glycolysis, in which a glucose molecule is broken down into two pyruvic acid molecules and some of its stored energy is released and is used in forming energy carriers, such as ATP molecule. (ATP, adenosine triphosphate, is the molecule that acts as a store and source of energy within cells. It releases a large amount of energy when it breaks down into smaller molecules; energy from food is used to reverse the process.)

2. Krebs cycle, in which pyruvate molecules are broken down in the presence of oxygen. Most of the energy is released during this stage. The overall chemical reaction (which is the reverse of photosynthesis) is:

$C_6H_{12}O_6 + 6O_2 = 6CO_2 + 6H_2O + energy$

Anaerobic respiration occurs without the presence of oxygen and therefore the Krebs cycle cannot take place. It produces only two ATP molecules compared to the 38 ATP molecules produced during aerobic respiration.

Fermentation is an anaerobic reaction. This process does not require oxygen.

Instead, glucose is broken down in ethanol (ethyl alcohol) and carbon dioxide. The overall reaction is:

$$C_6H_{12}O_6 = 2C_5H_5OH + 2CO_2 + energy$$

This reaction occurs also in some tissues of higher organisms when they are deprived of oxygen.

79. Science

'The intellectual and practical activity encompassing the systematic study of the structure and behaviour of the physical and natural world through observation and experiments' – that's how *Oxford Dictionary* defines science. To English biologist Thomas Huxley, 'Science is simply common sense at its best; that is, rigidly accurate in observation, and merciless to fallacy in logic.' Nobel-Prize-winning physicist Richard Feynman says humorously, 'As a matter of fact, I can also define science another way: Science is the belief in the ignorance of experts.' Put simply, science is both knowledge and process, and its purpose is to produces useful models of reality.

The following definitions are accepted by scientists, and are common to most branches of science:

- *Pure science (basic research)*: Investigation of nature by experiment or observation in an attempt to satisfy the need to know.
- *Applied science*: The application of scientific knowledge and principles to some perceived practical purpose.
- *Technology*: Application of scientific discoveries and inventions for industrial purposes.

  For example, a biochemist is a pure scientist when studying the properties of biochemical compounds. He or she is an applied scientist when exploring the physiological effects of some new drug. If he or she manufactures this drug by an industrial process this biochemist is a technologist.
- *System* is part of the material world which scientists select for study and experimentation. For example, astronomers study stars and the solar system; biologists study living things; and geologists study rocks and minerals.

- Sometimes scientists use a *model* –a mathematical or visual picture of a particular set of phenomena – to study the behaviour of a system, such as climate. A model may be mathematical or physical. A mathematical model consists of equations and step-by-step rules that reflect what happens in real event. A physical model represents a real object. A model is never perfect and scientists continually update their models on the basis or new observations.
- The only way to study a large population (people or things) is to select a representative *sample*. There are many techniques to collect random samples, but each may lead to some error.

80. Scientific Method

The scientific method is a continuous interplay of observation and hypothesis: observations lead to new hypotheses, which guide more experiments, which help to change existing theories. The method involves the following sequence:

(1) Observations and search for data.
(2) Hypothesis to explain observations.
(3) Experiments to test hypothesis.
(4) Formulation of theory.
(5) Experimental confirmation of theory.
(6) Mathematical or empirical confirmation of theory into scientific law.
(7) Use of scientific law to predict behaviour of nature.

This process is continuous as Einstein said, 'No amount of experimentation can ever prove me right; a single experiment can prove me wrong.'

It is the *modus operandi* of science – rather than facts – that makes science unique. Here're some useful terms:

- *Empirical*: Refers to a formula, equation, graph, analysis or relationship based on observation and experiments. An empirical relationship is not necessarily supported by scientific theories.
- *Induction*: The logical process used to build a general statement from a series of related observations. Hypothesis is a stage beyond induction.
- *Hypothesis* is a tentative explanation of observed facts. A hypothesis is assumed to be tenable for purposes of investigation. Every theory or law in science begins as a hypothesis. A hypothesis can be confirmed by experiments, which are observations under controlled conditions. When observations or experimental data do not fit in the hypothesis, it must be

changed or discarded.

- An *experiment* is observation under controlled conditions. When observations or experimental data do not fit the hypothesis, it must be changed or discarded.
- *Theory:* A theory is a hypothesis that has been tested by experiments, and to which no exceptions have been found.
- *Scientific law:* A theory that has received mathematical verification
- *Deduction:* A logical process in which a theory or law is used to generate specific information. For example, from his laws of motion and Kepler's laws Newton deduced that an inverse square centripetal force attracts Earth and its moon to the centre of their orbits (that is, the force is inversely proportional to the square of distance between Earth and the moon).
- *Paradox:* A proposition that seems absurd or self-contradictory, but is or may be true.
- *Rule:* A set of directions concerning method or procedure.
- *Postulate or axiom:* A generally accepted principle or proposition.

81. Scientific Notation

Very large and very small numbers are inconvenient to read and difficult to compare. To overcome this difficulty, we use scientific notation. This notation has many advantages over other methods and is commonly used in scientific calculators and by scientists, mathematicians and engineers. In scientific notation – also known as powers-of-10 notation –a number is expressed in the form Nx10^m, where N is a number between 1 and 10 and m is the appropriate power (exponent). For example, 69800000 becomes 6.98x10^7 and 0.00000698 becomes 6.98x10^{-6} (exponent is the number of places the decimal point shifts to give the original number).

As shown in the table below, a **positive** exponent show that the decimal point is shifted that number of places to the right. A **negative** exponent show that the decimal point is shifted that number of places to the left.

Number	Scientific notation
1	1×10^0
10	1×10^1
100	1×10^2
1000	1×10^3
10 000	1×10^4
0.1	1×10^{-1}
0.01	1×10^{-2}
0.001	1×10^{-3}
0.000 1	1×10^{-4}
0.000 01	1×10^{-5}

Significant numbers

The accurately known digits in a calculated number are known as significant figures. For example, 672 584, 45632 69, and 0.005 209 27 reduced to four significant figures become 672 600, 45.63, and 0.005 209. The result of a calculation never has more significant figures than the input data. If, for example, in a calculation $\pi = 31.4$, then the answer cannot have more than three significant figures. The rules for writing significant figures are:

- All nonzero digits are significant: 24.5 has three significant figures.
- All zero and nonzero digits in the base number of scientific notations are significant:
  - $\times 10^2$ has three significant figures.
- In numbers less than 1, zeros on the right of a decimal point are not significant: 0.004 has one significant figure.
- When a zero appears between two nonzero numbers, it is significant: 10.2 as three significant numbers.
- Zeros to the right of a decimal point and to the right of a nonzero digit are significant: 34.230 has five significant figures.

Order of magnitude

An approximation to the nearest power of 10 a number is called its order of magnitude. For example, a dollar ($= 100 \, c = 10^2 \, c$) is two order of magnitude more valuable than one cent. Orders of magnitude provide an easy way to compare measurements. For example, when rounded off to the nearest power of 10, the orders of magnitude of the masses of Earth are 10^{25} and 10^{-30} kg respectively, or Earth is 55 powers of magnitude more massive than the electron.

82. Sleep

Sleep is a state of reduced awareness and activity. The onset of sleep is marked by changes in the electrical activity of the brain, which can be recorded by an electroencephalograph (it records electrical impulses generated by the brain by hooking up electrodes to the head). During sleep the metabolic rate falls about 15 per cent: most of the muscles relax, and blood pressure, body temperature, heat rate and breathing are decreased. It's a misunderstanding that the sleeping brain isn't doing anything: it's busy organising memories and picking out the most important information – making you come up with new ideas.

We spend nearly one-thirds of our lives asleep. We are not alone; all animals sleep in one form or another. Body size appears to be related to the amount of sleep a species need. The larger the animal, less sleep it needs: elephant (about 3 hours), humans (8 hours), dog (10 hours) and cat (12 hours), platypus (14 hours). The reason perhaps is that neurons in the brains of smaller animals are more prone to injury because of higher metabolic rate. Consequently, their neurons require more time for strengthening connections with their partners.

Studies of the brain during sleep show that it is highly active while the body is passive. The active brain circuits produce electrical impulses, or waves, which can be recorded by an electroencephalograph (EEG). The recordings – and the measurement of the eyes and the limbs – show five different stages of sleep:

Awake: Sleep-on neurons, a small group of neurons in the forebrain responsible for inducing sleep, are inactive; alpha brainwaves (relaxation)

Stage 1: Marks the transition between awake and asleep; sleep-on neurons fire; shallow brainwaves; muscles relax and eye movement slows down

Stage 2: It lasts the longest; bursts of wave activity; sleep talking

Stages 3 and 4: It last for about 30 minutes; deepest, or slow-wave, sleep; delta waves appear; sleepwalking and bedwetting

Stage 5 or rapid eye movement (REM) sleep: Lasts 10 to 15 minutes; accompanied by rapid, jerky eye movements; heart rate, blood pressure and body temperature become much more variable; the brain is highly active (it's on fire); vivid dreams occur (dreams also occur in other stages but they are not vivid)

During the night, these stages of slow-wave and REM sleep are repeated in roughly 90-minutes cycles until waking occurs. There is more slow-wave sleep early on and more REM sleep towards morning.

Research shows that learning continues in sleep. Sleep not only helps retain newly learned information but may also assist in recalling it when you need it.

A new piece of information such as a new phone number or a new name would be quickly forgotten unless it becomes a permanent record in our long-term memory. Conversion from short-term memory (lasting less than 30 seconds) to long-term memory (lasting longer than 30 seconds) is called 'memory consolidation'. It occurs when connections between neurons as well as different brain regions are strengthened. Changes in neurons typically take place within the first minutes or hours of learning. Other changes, such as the reorganisation of neuron networks that handle individual memories, can take several days or years.

Sleep not only makes memories stronger, it also reorganises and restructures memories to help you produce new and creative ideas. A sleepless night, on the other hand, results in poor subsequent retention of information. Staying up all night to cram for an exam may work for you, but most likely you will forget what you have learned in a few days.

83. Solar System

The solar system is a 4600-million-year-old star – the Sun – and its planets in one of the spiral arms of the Milky Way, about 26,000 light years from the galaxy's centre. It has eight planets – (in order of increasing distance from the Sun) Mercury, Venus, Earth, Mars, Jupiter, Saturn, Uranus and Neptune – dwarf planets (Pluto, Eris, Ceres and others), 140 known natural satellite, and smaller bodies such as the asteroids, the comets and meteoroids. The solar system moves as a whole in an approximate circular orbit around the centre of the Milky Way, taking about 220 years to complete the orbit.

The standard theory of the origin of the solar system is that about 4.6 billion years ago a gas and dust cloud, perhaps the debris of a supernova explosion, began to collapse under its own gravity to produce a large rotating sphere within a disc of gas. Gravity compressed the sphere, creating enormous heat that triggered fusion of hydrogen atoms. The sphere became our Sun (which is still simply a ball of burning hydrogen); the encircling gas disc cooled and condensed, forming the planets.

The planets and their characteristics are described in 'Planets'. Below is a brief description of other solar system bodies.

Meteorites are chunks of extraterrestrial matter, remnants of geological processes that formed our solar system. When these chunks enter the Earth's atmosphere they shine brightly because of the heat produced by friction with the air. Most chunks are too small – usually the size of a grain of sand, but no larger than a pea – to survive the trip, and are called **meteors** (or falling stars or shooting stars because they leave momentary streaks of light in the sky). Very rarely, a large chunk, which flashes like a fireball in the sky, survives its journey through the air to hit the ground. The falling object – a solid piece of stone or iron, often weighing many kilograms – is known as a meteorite. (Until a meteor or a meteorite enters the Earth's atmosphere it is known as a *meteoroid*.)

Asteroids are small bodies orbiting the Sun, mostly in between Mars and Jupiter. These pockmarked giant peanut-like rocks are in fact leftovers from the formation of the planets. The largest three asteroids are Ceres, Pallas and Vesta, with average diameters 930, 520 and 500 kilometres respectively. About 200 asteroids are larger than 100 kilometres across; 800 larger than 30 kilometres. About a million are 1 kilometre or more in diameter; and billions are of boulder or pebble size. Most of the asteroids orbit within a vast, doughnut-shaped ring between Mars and Jupiter, known as the main belt. Occasionally, a collision may kick an asteroid out of the belt, sending it onto a dangerous path that crosses Earth's orbit. These stray asteroids take up an orbit that loops past Earth, and are called 'Earth-crossers'. This knowledge frightens astronomers. What if one of them comes too close to Earth? What cataclysm would such a rogue rock cause if it slams into Earth? The number of asteroids is very large, but the space they occupy is enormous. Most asteroids stay millions of kilometres apart. It's not like *Star Wars* or *Star Trek* spaceships weaving their way through flying rocks. But real collisions are possible with a spacecraft or Spaceship Earth.

A *comet* is a unique cosmic phenomenon: it suddenly appears in the sky, it blazes for a few days, it wows earthlings, it disappears. Comets are independent masses of ice and dust – chunks of matter left over from the birth of the solar system – that orbit the Sun. There are somewhere between 2 trillion and 5 trillion comets that circle the solar system in a halo-like cloud – the Oort cloud – between 20,000 and 100,000 astronomical units from the Sun (one astronomical unit is the distance between the Earth and Sun, about 150 million kilometres).

84. Solution

A solution is a homogenous mixture. It consists of solute, the dissolved substance, and the solvent, the substance in which it is dissolved. The solute and the solvent may be solid, liquid or gas. The solution has the same physical state as the solvent. A solution containing a small amount of solute is termed dilute. When a large amount of solute is dissolved the solution is called concentrated. Dilute and concentrated are qualitative terms and do not give information about the exact amount of solute present in a solution. Quantitative methods are used to express exact concentration of a solution.

Because the solute and the solvent may be solid, liquid or gas, nine types of solutions are possible:

Solute	Solvent	Example
solid	solid	sterling silver (copper in silver)
liquid	sold	amalgams (mercury in silver)
gas	solid	hydrogen in palladium metal
solid	liquid	brine (salt in water)
liquid	liquid	alcohol in water
gas	liquid	oxygen in water
solid	gas	odour of moth balls (naphthalene in air)
liquid	gas	humid air (water in air)
gas	gas	air (oxygen and other gases dissolved in nitrogen)

When no more solute can be dissolved in a solvent the solution is said to be saturated. A solution containing less solute than it is capable of dissolving at a given temperature is called unsaturated. Under certain conditions a solution may have an excess of solute. Such a solution is called supersaturated. The concentration of a saturated solution is described as its solubility.

A true solution consists of dissolved ions and molecules. If a solution is made up of bigger particles, it's called colloid. Colloids play an important part in our lives. Milk, butter and cheese, dough and bread, and the yolk and white of egg are all colloids.

85. Sound

A source of sound is a vibrating object which causes a train of compressions (high pressures) and rarefactions (low pressures) to move through a medium. We can only hear sound between frequencies in the range of from about 20 hertz to about 20,000 hertz. This is called the audible range. Different frequencies sound different to the ear. High-frequency vibrations produce high-pitch notes; low-frequency vibrations low-pitch notes. Loudness – measured in decibel (symbol dB) – is our own perception of sound. A 10-dB increase doubles the loudness of sound: a sound of 60 dB sounds twice as loud as sound of 50 dB.

Sound waves with frequencies above 20,000 hertz are called ultrasound. Ultrasound waves, like all sound waves, consist of cycles of compression and rarefaction. Ultrasound has many important applications in industry and medicine. In industry it is used for cleaning, for drilling holes, for measuring electrical signals, detecting gas leaks and for burglar alarms. In medicine it is used for observing foetuses and internal organs.

Sound waves with frequencies below 20 hertz are called infrasound. It is generated by many natural and industrial processes, including wind, car engines and blast furnaces. We cannot hear ultrasound but it is suspected of causing nausea, dizziness and headaches in many people.

Sound needs a medium to travel and its speed is different in different materials. The speed depends on temperature, but this is significant for gases only. At 20°C, the speed of sound (in metres per second) in air is 343, in water 1440, in gases about 4500, and in steel about 5000.

The pitch of a sound refers to whether it is high like the sound of a violin, or low like the sound of a bass drum. It is determined mainly by frequency, but also depends on loudness – the intensity of sound to the human ear. A high-pitched sound has a high frequency; a low-pitched sound, low frequency.

Pitch		Frequency
high	upper limit of hearing	20,000 Hz (hertz)
	whistle	10,000 Hz
	high note (soprano)	1000 Hz
	low note (bass)	100 Hz
low	Lower limit of hearing	20 Hz

Supersonic speed

It is a speed greater than the speed of sound, which is about 1200 kilometres per hour. An aircraft travelling at supersonic speed causes shock waves containing a tremendous amount of sound energy. These shock waves produce a loud noise, known as sonic boom. Mach number is the ration between the sped of a body and the local speed of sound. Above 1 Mach the speed is supersonic.

Doppler Effect

Any source of sound or light moving away from an observer changes in frequency with reference to the observer. For example, the pitch of the whistle of a train changes as it runs past a person standing on a platform: it is higher when the train is approaching the person; and lower when it is moving away from the person.

In 1842 Austrian scientist Christian Johann Doppler explained this phenomenon by pointing out that when the source of sound is moving towards the observer, sound waves reach the ear at shorter intervals, hence the higher pitch. When the source is moving away, the waves reach the ear at longer intervals, hence the lower pitch. The Doppler Effect also occurs when the source of sound is stationary and the observer is moving.

In 1848 it was shown that the Doppler Effect also applies to light coming from distant stars.

86. Space

Classical geometry (or Euclidean) deals with flat space, but we live on a curved surface – planet Earth which inhabits a curved universe. In his theory of general relativity, Einstein described three main ideas: (1) Space and matter are not rigid; their form and structure is influenced by matter and energy. (2) Matter and energy determine how space and space-time (three dimensions of space and the fourth dimension of time) curve. (3) Space and its curvature determine how matter moves. This does not mean that classical geometry is wrong. Euclidean geometry works accurately as long as the curvature is very small.

The 19th-century German mathematician Bernhard Riemann was one of the first mathematicians to study curved space. His work created a new geometry now known as Riemann geometry or elliptical geometry

In Riemann geometry there are no true parallel lines, all straight lines are equal in length, and the sum of angles of a triangle is always greater than 180 degrees. The last idea allows all longitudinal lines to cross at both the North and South poles.

Einstein was highly impressed by Riemann's idea of curved space and applied it in his general theory of relativity.

Relation between space and matter is easy to understand if you imagine a weight placed on a rubber sheet. The weight will cause the sheet to bend. Similarly, matter stretches the space-time around it. If the mass of a body, say a massive star, is highly concentrated, the curvature of space-time it produces becomes infinite. At this end stage of the curvature of space-time, space is so curved that once matter and energy enters that space, it can never get out. This is the 'black hole', a point of infinite density where mass has no volume and both time and space stops. Nothing – not even light – can escape a black hole. At first it was thought that black holes were an abstract mathematical idea, but it is now well known that black holes actually exist, just as Einstein's theory predicts.

The noted 20th-century British astrophysicist Arthur Eddington explains how geometers and physicists see the curved space:

> To the pure geometer the radius of curvature is an incidental characteristic – like the grin of the Cheshire cat. To the physicist it is an indistinguishable characteristic. It would be going too far to say that to the physicist the cat is merely incidental to the grin. Physics is concerned with the interrelatedness such as the interrelatedness of cats and grins. In this case the 'cat without a grin' and the 'grin without a cat' are equally set aside as purely mathematical phantasies. – *The Expanding Universe* (1933)

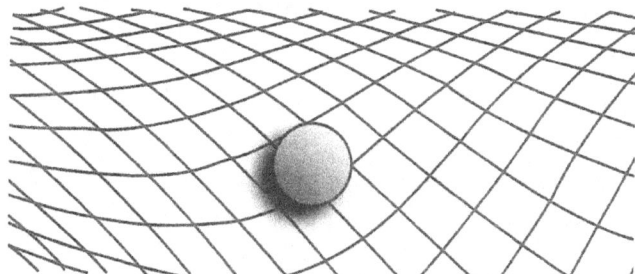

87. Species

A species is a population of organisms consisting of similar individuals capable of breeding among themselves. The term is used for both living and extinct organisms. It is a unit of classification of plants and animals. All species are given a two-word scientific name (in Latin) – we are *Homo sapiens*. The first word is the genus name and begins with a capital letter; the second is the species name which begins with a lower-case letter. So far about 1.7 million species have been identified. The estimate for the total number of species on Earth ranges from 10 to 30 million.

In 1686 English scientist John Ray was the first to use the word species (the Latin for 'kind' or 'form') in its current scientific sense. His concept of species was based on the study of 18,600 species of plants. He said; 'After a long and considerable investigation, no surer criteria for determining species has occurred to me than the distinguishing features that perpetuate themselves in propagation from seed. Thus, no matter what variations occur in the individual or the species, if they spring from the seed of the one and the same plant, they are accidental variations and not such as to distinguish a species.'

Swedish natural scientist Carl von Linn, known by his Latinised name Linnaeus, was the first to describe and classify species. In 1735 he published his famous book, *Systema Naturae*, which recorded some 9000 species of plants and animals. In this book he introduced the two-word system for naming species. The Linnaean system provides a concise, orderly method for classification. The system is still used widely but now the genetic code of an organism provides a better method for classification.

Of the 1.7 million know species, at least two-thirds occur in tropical forests. Only about one-fifth live in the sea. Insects account for the more than half of all known species. Of the insects, beetles are the most diverse: around one in five of all recorded species of organisms is a beetle. This fact prompted the eminent

Scottish geneticist J.B.S. Haldane to comment that the Creator has 'an inordinate fondness for beetles'.

Why is it important to know the total number of species worldwide? As a large number of species are becoming extinct every year, knowing the total number and distribution of species is essential for developing an effective program to conserve as much as possible of the remaining biodiversity on our planet.

88. Star

Stars are simply balls of burning gases that are held together by gravity. A perfect example of elegance in simplicity. The Sun burns hydrogen to helium. The process, which occurs during most of every star's life, gives off energy that is radiated from the surface of the star as heat and light. The universe has ten times as many stars as grains of sand on Earth – 100,000 billion billion stars, to be precise. To us all stars look similar, but no two stars are the same. Astronomers classify the stars in our galaxy by their luminosity colour, size and age.

To us, stars also appear changeless. But stars are born, they live for millions of years, and they die.

The birth sites of stars are the dark clouds of gas and dust in our galaxy. The clouds, which are clumps of hydrogen atoms with a sprinkling of helium, are not uniform; they contain regions differing by density (1,000 to 10 million molecules per cubic centimetre) and temperature (–263 to –173°C), and regions with shapes ranging from spheroids to elongated tubes. Gravity tries to pull these clouds into the smallest possible space. Compression causes the gas to become hotter. Eventually the temperature and pressure rise high enough to ignite the gas. Hydrogen starts turning into helium, which creates vast amounts of energy. A star is born.

At the extremely high temperature in the interior of a star, atomic nuclei are stripped of their electrons, moving freely among the electrons themselves. Occasionally two hydrogen nuclei (two protons) combine to form a deuterium nucleus (one proton, one neutron), producing energy in the process. When another proton collides with a deuterium nucleus, they combine to form helium-3 (two protons, one neutron). Finally, two helium-3 nuclei combine to form stable helium-4 (two protons, two neutrons). These types of reactions in which lighter nuclei combine to form a heavier nucleus are called nuclear fusions (a nuclear fission, on the other hand, is the breaking up of a heavy nucleus into two or more lighter nuclei). The nuclear fusion produces energy that is give off in all forms of electromagnetic radiation – infrared rays, light, X-rays, radio waves, etc.

Most of the elements higher than hydrogen in the universe are created, or synthesised, in stars when lighter nuclei fuse to make heavier nuclei. This process is called nucleosynthesis. After a star exhausts its supply of hydrogen, it burns helium to form beryllium, carbon and oxygen. When the star exhausts it supply of helium, it shrinks and its temperature rises to 1000 million degrees. The rising temperature triggers a new series of reactions in which carbon, oxygen and other elements combine to form iron and nickel. When the star has burned everything into iron and nickel, it explodes into a supernova. The elements higher than nickel are formed during supernova explosions.

89. Stem Cell

Stem cells are a class of cells that have the remarkable potential to develop into different types of cells in the body during early life and growth. Commonly, they come from two sources: (a) embryonic stem cells – derived from a four- or five-day-old human embryo; and (b) adult stem cells – found in tissues such as the brain, bone marrow, skin and the liver. Stems cells have three general properties: (a) they can divide and renew themselves for long periods; (b) they are unspecialised; and (c) they have the potential to give rise to specialised cell types –skin, muscle, bone etc.

Every cell in the body is derived from first few stem cells formed in the early stages of the development of embryo. As they have the capacity to serve any function after they are manipulated to specialise, they offer new potentials for treating diseases such as Alzheimer's, diabetes, heart disease, spinal cord injury. For example, experiments on mice and other animals show that bone marrow cells transplanted into a damaged heart stimulate the regeneration of damaged heart muscle cells. Stem cell research can also help in the development of new drugs by testing a new drug on tissues grown from stem cells rather than testing it on humans.

Stem-cell research is still at a very early stage, and only a small number of trials on humans have been conducted. Research on human embryonic stem cells is highly controversial as it requires the destruction of a fertilised egg. Those who oppose stem-cell research on ethical grounds ask why the fertilised egg wasn't given the chance to develop into a fully-developed human. They believe that every embryo has the same right to live and grow as any other human.

British scientist John Gurdon launched the field of stem cell science in 1962 when he discovered that the DNA code in the nucleus of an adult frog held all the information to develop into every type of cell. He then successfully cloned tadpoles. In 2006, Japanese scientist Shinya Yamanaka discovered that mature skin cells in mice can be reprogrammed to become embryo-like cells again. This means stem cells need not to be taken from embryos. Yamanaka's work allays the fears of opponents of stem cell research.

In 2012, Gurdon and Yamanaka were awarded Nobel Prize for physiology or medicine for their pioneering work on stem cells. When Gurdon was a 15-year-old schoolboy at Eton, his science teacher dismissed his ambition of becoming a scientist as 'quite ridiculous'. The moral of the story: follow your dreams.

90. Symbiosis

Symbiosis (from a Greek word for 'living together') is a close ecological relationship between different kinds of organisms of two or more different species. It refers to organisms that live in close proximity; often one cannot live without the other. Sometimes one organism (for example, tapeworm) actually lives inside the other organism (human intestine). The term is usually used for associations in which both species benefit (*mutualism*), but it may be used for associations in which at least one species benefits while the other species is neither benefited nor harmed (*commensalism*), or one species benefit, the other is harmed (*parasitism*).

In *mutualism*, both species benefit. Examples include: (a) nitrogen-fixing bacteria, Rhizobium, in the roots of legumes (they enrich the soil with nitrogen compounds); and (b) zooxanthellae algae in a coral animal (zooxanthellae uses photosynthesis to provide the coral with sugars in return for nitrogen and other nutrients from the coral).

In *commensalism*, only one species benefit. An example is plants that grow on tree trunks.

In *parasitism*, one species (the parasite) receives benefit at the expense of other species (the host). A parasite lives in or on another organism and obtains food from it. Parasites such as tapeworm, which live inside the body of the host, are called endoparasites. Parasites such as fleas or lice, which live on the outside of the host, are called ectoparasites. Some parasites can be employed to reduce populations of animals and pests.

Biological pest control – use of biological means to decrease the population of pests – relies on parasitism, predation (a predator is an organism that feeds on other organisms which are called the prey – predator is not a parasite) or other natural means. Biological control is preferable to chemical control because it does not involve using pesticides that affect other organisms in an ecosystem. One of the world's most successful effort in biological control is the eradication of prickly pear (*Opuntia stricta*), a shrub-size cactus plant introduced to Australia as a garden plant during the 1880s, which became a pest though many areas of eastern Australia. By the 1920s millions of hectares of agricultural land were infested. In 1926 Australin scientists introduced the moth *Cactoblastis cactorum*, a native of Argentina, whose caterpillars' tunnel in the stems of prickly pear, gutting and killing it. By 1934 *Cactoblastis* virtually eradicated the noxious weed. The relation between *Cactoblastis cactorum* and *Opuntia* cactus species is parasitic as the moth feeds on the host cactus.

91. Systems

A system is a collection of interacting and interdependent things. The things – or parts or components – of the system, can be almost anything: objects, organisms, organisations, machines, processes, ideas or numbers. The definition of a system – whether it is a chemical system, an ecosystem or the solar system – is incomplete until we include enough parts that are essential to the understanding of their relationship with each other. For example, we can explain tides in terms of a system that includes only Earth and the Moon. But an accurate system would also include the Sun as its gravitational force influences Earth.

Let's look at systems in context of chemistry.

Chemistry involves the study of chemical systems and the changes which take place in them. A system is the part of the material world which the chemist selects for study and experimentation. Everything else in the environment is called the surroundings. For example, if you wish to study the behaviour of sugar in the presence of sulphuric acid, you will probably isolate them in a test tube (or any other container). The contents of the test tube are referred to as the system. Everything else around the contents, including the test tube, is referred to as surroundings.

A chemical system is said to be closed if the amount of matter in it remains constant. Al open system, on the other hand, is one in which, the amount of matter changes with time.

Most systems are not closed in the sense that they are not mutually exclusive. They may be closely related and it becomes difficult to draw boundaries that separate them. For example, these days it is impossible to think of the transport system separate from the communications system; they extensively interrelated

If you were interested in studying the chemical processes which take place in your body, then you body would be called the system. Similarly, the solar system becomes the system if you are interested in the chemistry of the formation of the planets. Any part of a system is considered as a system, or subsystem, with its own parts and interactions. For example, Mars is a system in itself with the solar system.

A system may be studied by using a model – a computer model, a plan, a drawing, a set of equations or simply a mental image – to study the behaviour of a system, such as climate. A model may be mathematical or physical. A mathematical model consists of equations and step-by-step rules that reflect what happens in real event. A physical model represents a real object.

Feedback helps to control the performance of a system by using its output to modify its input. For example, a thermostat in a gas heater works properly because it senses the temperature (a measure of output) of the system and feeds it back to the system to modify gas supply (input).

92. Technology

Technology is the application of scientific knowledge for practical purposes. The factories and industries of today, which supply daily needs and desires of billions of human beings, represent technology. It is technology that defines the cultural, social, economic, political and physical aspects of human existence. But it is likely that science will determine the future course of civilisation. For its progress technology depends upon science – but this dependence is not one-way process. Technological innovations also help in advancing scientific knowledge. Technology and science are interrelated and dependent on each other but they are not synonymous and their goals are different.

The invention of printing press – one of the most important technical breakthroughs in history – is a great example how technology defines our lives.

Movable type was invented in China (made from clay in *c.*1041; made from hard wood in *c.*1297). Gutenberg was a young engraver and gem-cutter when he thought of using movable type to compose whole books. He experimented over several years, borrowing large sums of money to cover costs. He made moulds of letters for casting individual characters in metal. He invented devices for composing the types on a wooden plate and for inking the composition evenly, and finally a hand-printing press for making impressions of the plates on paper. He tested his press by printing an old German poem over and over. He was now ready for the job he'd been dreaming of for years, printing the whole Bible in Latin. He composed 1,283 pages of 42 lines each, and printed 180 copies. The first book ever printed from movable type was ready in 1455.

In Europe in the early 14th century there were only handwritten books or some short texts, printed using wooden blocks into which words had been carved. First time in history books could be reproduced en mass.

Gutenberg's printing press liberated knowledge from the preserve of the privileged few. The printing press made books cheap and accessible. Within a short time books, which were sold in gold guineas, were sold in copper pennies, making books accessible to everyone.

However, state and the church were fearful of the invention as they saw it as something that would undermine their authority. In England the king required all printers to seek approval before they could print anything. Up to this time the Bible was owned and interpreted by priests only. They were worried they would lose control over their parishioners. The easy availability of the Bible led to widespread criticism of the prevailing theological thoughts and influenced the development of schisms, which resulted in the Protestant Reformation led by the German monk Martin Luther.

However, printing presses were accepted readily and within years they were set all over Europe. Nobility and clergy were no longer the only repositories of knowledge. Knowledge became accessible. Now the internet has added a new dimension to accessibility of information.

Some new technologies

Biotechnology is the term used for technologies which use living organisms to manufacture useful chemicals. Some biotechnologies are traditional (fermentation, for example), but most are based on modern advances in molecular biology.

Nanotechnology is the technology that deals with objects less than 100 nanometres (a nanometre is millionth of a millimetre), especially manipulation of individual atoms and molecules.

Information technology is the application of appropriate technologies to solve transmit, store and manipulate information to solve problems and make decisions. At present information technology blends computing, data and telecommunications and digital electronics. Quantum computers, based on quantum mechanics, are not far away.

93. Temperature

Temperature is a measure of hotness or coldness of a body. It is thus a measure of the kinetic energy of individual molecules in the body. *Thermodynamic temperature* **is the absolute measure of temperature. Its starting point, or zero point, is** *absolute zero*, **which is the theoretical lowest limit of temperature. At this point all atomic motion (which is what constitutes heat) has ceased. Absolute zero cannot actually be reached, as to reach it an infinite amount of energy is required. No thermometer can be built to measure absolute zero, where molecules have the lowest possible energy and the entropy.**

Celsius (°C) and Fahrenheit (°F) are non-SI units of temperature and are used for everyday measurements (°F is used mainly in the US). Freezing and boiling points of water (at standard pressure; that is, 101.325 kilopascal) in these scales are 0°C/32°F and 110°C/212°F respectively.

In scientific work, *absolute temperature scale* is used. The triple point of a substance is the temperature and pressure at which the substance can exist in its three states at the same time – solid, liquid and gas. The triple point for water occurs at 0.01°C and 611.73 pascal. In 1848, British scientist William Thomson (Lord Kelvin) devised his own temperature scale based on the triple point of water. He used a 100-unit scale, like that in the Celsius scale, and by extrapolation arrived at –273° as the absolute zero point. In this scale, water freezes at 273° and boils at 373°.

The scale is now known as the kelvin scale (symbol K without the degree sign) in honour of Thomson, and the absolute zero (0 K) has been refined to –273.15 °C. Temperatures can be converted from one scale to another using the relationship K = °C + 273.15. Kelvin units are equal to Celsius degrees.

Zero-point energy is the energy possessed by a substance at absolute zero. According to the uncertainty principle, atoms and molecules can only exist in certain energy levels: the lowest energy level is called the ground state and all higher levels are called excited states. At absolute zero all particles are in the ground state.

Temperature conversion factors

°C to K	°C + 273.15
K to °C	K − 273.15
°C to °F	(°C × 1.8) + 32
°F to °C	(°F − 32) ÷ 1.8

94. Theory

In science, a theory is a hypothesis – tentative explanation of observed facts – that has been tested by experiments, and to which no exceptions have been found. A scientific theory can be used to predict phenomena. For example, Einstein's general theory of relativity – gravity is not a force but warping of space-time – predicted that starlight passing near a massive body like the Sun would be bent. In 1919, British astronomer Arthur Eddington photographed a total solar eclipse off the coast of West Africa. The pictures showed that the starlight was deflected by the Sun's gravity, just as the theory had predicted.

Once a scientific theory has received mathematical verification it's known as a scientific law. A law, such as Newton's law of gravitation, is a concise and general statement about how nature behaves and brings unity to many observations. For less general statements, such as Archimedes principle, the term scientific principle is used.

Theories are by nature speculative and not certain. They are never fixed, but continually changing. Here's is a brief account of major events in the development of the theory of gravity.

The 2nd-century BC Greek mathematician and geographer Claudius Ptolemy synthesised the work of his predecessors in his famous book, *Almagest* (Arabic for 'greatest'). In this book he said that the earth was spherical in shape; that it is was situated at the centre of the universe; and it didn't move at all. This erroneous idea dominated astronomical thought for 14 centuries until 1543 when Polish astronomer Nicolaus Copernicus declared that the Sun is at the centre of the solar system, fixed and immovable, and planets orbit around it in perfect circles.

In the early 17th-century German astronomer Johannes Kepler formulated his three laws of planetary motions. The first law said that the planets move in elliptical orbits with the Sun at one focus. Modern measurements show that planets do not precisely follow Kepler's laws; however, their development is considered a major landmark in the history of science. A few decades later Italian scientist Galileo investigated the laws of motion, experimented on falling bodies and formulated theories of planetary motion. Galileo's rules for falling bodies – all bodies fall with the same motion, and the motion is one of constant acceleration – formed the foundation on which English scientist Isaac Newton based his laws of motion. The publication of Newton's *Principia* in 1687 is considered one of the most important events in the history of science. In this book Newton published his famous law of universal gravitation: any two bodies attract each other with a force proportional to the product of their masses and inversely proportional to the square of the distance between them.

The universality of the law of gravitation was challenged in 1915 when Albert Einstein published his theory of general relativity: objects do not attract each other by exerting pull, but the presence of matter in space causes space to curve in such a manner that a gravitational field is set up. Gravity is the property of space itself.

The theory of general relativity is incompatible with quantum mechanics which explains the behaviour of matter in terms of elementary particles. Will there be a theory of quantum gravity?

Einstein once said that it's the theory that decides what we can observe. His theory of general relativity led astronomers to observe the total eclipse of 29 May 1919, which proved his theory.

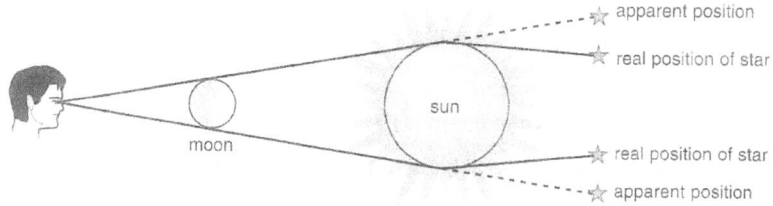

95. Time

Time is a mysterious concept; nonetheless, most fundamental of all concepts. The concept of time has fascinated scientists and poets alike. Poets use many metaphors to describe time but the most common one is the image of time as a river which flows forever. 'Time, dark time, secret time, forever flowing like a river,' wondered American novelist Thomas Wolfe. However, scientists have a more complex concept of time and that concept comes from Einstein who said that time was not an absolute quantity: space and time were interrelated and within certain limits the motions of space and time were interchangeable.

Until the beginning of the 20th century the scientists' concept of time was based on the ideas of Newton who believed that there is only one system of time 'flowing' through the universe. 'Absolute Space, in its own nature, without regard to anything external remains always similar and immovable ... Absolute, True and Mathematical Time, of itself, and from its own nature, flows equably without regard to anything external,' he said in 1686. This is the system of absolute time which implies that each time interval, such as a second or a year, is the same throughout the universe.

The Newtonians believed that like time, space is also absolute and it is spread throughout the universe and separates objects. In 1905 Einstein rejected these ideas. His explanation defied common sense, but to Einstein common sense was a 'deposit of prejudice laid down in the mind prior to the age of eighteen.' He once remarked that it was common sense that was once rejecting to the idea that Earth is round.

Einstein proposed the following postulates which are the basis of his theory of special relativity:

- The speed of light (in empty space) is the same for all observers no matter what their speed or what the speed of the source.

- The basis laws of nature should be the same no matter in what system or reference we are studying them.

He explained that space and time were interrelated; and within certain limits the motions of space and time were interchangeable. He also said that time was not an absolute quantity – it was like space, a matter of relative motion. In other words, our measurements of time are affected by our motion. The rate at which the clocks run depends upon their relative motion. A person running away from a clock would observe it to move more slowly than his own clock.

Theoretically, a spaceship travelling nearly at the speed of light (about 300,000 kilometres per second) would take nine years to make a return trip to Alpha Centauri, nearest star after the Sun; but because of for relativistic time changes on return to Earth the crew would find that many decades have gone by. However, the crew notice no change on the spaceship. From their point of view, the spaceship is stationery and Earth is moving at almost the speed of light and time on Earth slows down.

Relative time poses an interesting paradox: If one twin goes on a high-speed space journey, he or she would return younger than his or her sibling who stayed at home.

As long as Einstein's theory is supreme, time travel into space would remain in the realm of science fiction.

96. Universe

A 13.7-billion-year-old expanse of flat space filled with hundreds of billions of galaxies and vast clouds of glowing gas. This infinite space has been expanding and the galaxies are moving further apart. The rate of expansion is mind-boggling: imagine a pea growing to the size of the Milky Way in less time than it takes to blink. Only a tiny proportion of the universe's stuff is visible. The rest is known only as dark matter and dark energy. Their true nature is still a mystery. Clues to the destiny of the universe are hidden in dark matter and dark energy.

In 1929 American astronomer Edwin Hubble suggested that galaxies are moving away from us and each other at an ever-increasing rate. More distant the galaxy, the faster it is moving from us. This means that the universe is expanding like a balloon.

This is now known as Hubble's law. It shows that the ratio of velocity of galaxies to their distances is a constant. This constant is known as the Hubble constant, which is the present rate of expansion of the universe – 22 kilometres per second per million light years. Hubble's law presupposes that universe began in a Big Bang which took place about 13.7 billion years ago.

'The discovery that the universe is expanding was one of the great intellectual revolutions of the twentieth century,' remarks celebrated physicist Stephen Hawking in his mega-selling book *A Brief History of Time*.

The expanding universe also explains a deceptively simple question – why is the sky dark at night? – which has puzzled astronomers for centuries.

The distant galaxies are moving away from us at a speed so high that it diminishes the intensity of light we receive from them. In addition, this light shifts slightly towards the red end of the spectrum. Red light has less energy than blue light. These two effects significantly reduce the light we receive from distant galaxies, leaving only the nearby stars which we see as points of light in a darkened sky. However, in an infinite and stationary universe, uniformly filled with stars, our line of sight would always end at the surface of a star, and the entire sky should therefore be bright.

Accepting that the universe is expanding you might wonder whether it will go on expanding forever or whether the receding galaxies will someday stop and reverse their motion eventually falling together in a great collapse – a Big Crunch, a kind of reverse Big Bang? All recent observations show that the universe will expand forever.

There exist more matter in the universe than we see as bright galaxies and bright stars. This so-called 'missing mass' comprises up to 95 per cent mass of the universe and it is in the form of dark energy and dark matter. Roughly 70 per cent of the universe is dark energy; dark matter makes up about 25 per cent. Scientists are still in dark, pardon the pun, about dark energy and dark matter.

97. Vacuum

A vacuum is a space that contains no matter. No place in the universe has an absolute vacuum. Interstellar space comes very close, but it still has a few hydrogen atoms per cubic metre. It is impossible to create an absolute vacuum in a laboratory, as it has to be a region without any particles. There is always the chance that a neutrino would enter the region. Neutrinos – elementary particles with no charge and nearly no mass – rarely interact with matter. They are everywhere; trillions of them pass through our bodies each second yet we can't see or feel them.

In 4th century BC, Greek philosopher Aristotle said that 'nature abhors a vacuum'. This thought persisted for nearly two millenniums until scientists in the seventeenth century started to study the true nature of vacuum. One of them was Otto von Guericke, a wealthy amateur scientist in Germany, who in 1650 made a great technical advance by inventing a vacuum pump (sometimes called an air pump), a device to mechanically remove air from a container.

In 1651 Guericke, who was then the mayor of Magdeburg, performed a spectacular experiment before Emperor Ferdinand III and his courtiers. He used his vacuum pump to evacuate air from two brass hemispheres about 45-centimetres in diameter, with edges that fitted exactly. Eight horses were harnessed to one cup and eight to the other. But the team of sixteen horses could not pull the hemispheres apart. To amaze his audience further, he turned the tap, and as the air rushed into the sphere, he effortlessly pulled the cups apart.

Guericke not only used his vacuum pump to demonstrate the power of atmospheric pressure, he also used it to show that light, but not sound, can travel through a vacuum. And one cannot light a candle in a vacuum.

In 1662 English chemist Robert Boyle made an efficient vacuum pump which he used to establish his law, now familiar to all schoolchildren. He also used his pump to experiment on respiration and combustion and showed that air was necessary for life as well as for burning.

Vacuum is a condition of standard atmospheric pressure (101.325 kilopascal, kPa). The quality of 'partial vacuum' relates to how close it is to an absolute vacuum; it describes the region of space below 101.325 kPa. Gas pressure is due to the bombardment of the walls of the container by the moving molecules. As we remove gas from the container, there are less and less molecules bombarding the walls. A partial vacuum means less number of molecules and therefore a low pressure.

98. Virus

Viruses are disease-producing agents, and cause diseases in humans, plants and animals. A virus is very small (20 to 300 nanometres in diameter) and consists of a small number of genes, made of DNA or RNA, which are encased in a protective coat of protein. Many viruses are shaped in the form of icosahedrons, which has 20 triangular faces. Spikes protruding from its outer surface help the virus to recognise and attached to a cell. Viruses can grow and multiply only when inside living cells. They are able to direct the machinery of their host cells to manufacture more viruses.

Human immunodeficiency virus (HIV) causes AIDS (acquired immune deficiency syndrome). HIV is carried only in body fluids – mainly semen, blood and vaginal secretions – and it cannot live long outside these fluids. It attacks the immune system, the body's defence against invading germs. As a result people with the disease are subject to a wide range of infections and some rare cancers. It can take years for symptoms to appear in an affected person. The disease was not recognised until 1981. Studies of medical records and tissue samples suggest that a few cases of disease may have occurred in the West as early as 1959. As yet there is no vaccine or cure, although suitable drug therapy can prevent damage to the immune system and stop or delay the progress of the symptoms of the disease.

HIV virus is a retrovirus. A *retrovirus* is a virus with its genetic information in the form of a single strand of RNA. It contains an enzyme called reverse transcriptase which enables it to make DNA when inside a host cell. The viral DNA joins the host cell's DNA and becomes part of the body's genetic information code. Viruses are not organisms and are not included in the classification of organisms. Although they possess some characteristics of living things, they lack many features essential for life.

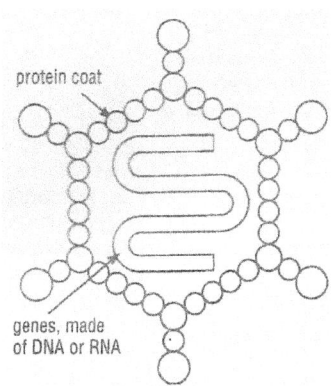

protein coat

genes, made
of DNA or RNA

99. Water

Water is a miracle that makes our world possible. The story of life as we know is the story of water and carbon. Without them there would be no life on our planet, but could there be life without water or carbon (or both) on other planets? That we do not know – yet. The human body contains about 70 per cent of water by mass. In some organisms up to 90 per cent of their body weight comes from water. Pure water is colourless, tasteless and odourless. Natural waters contain dissolved gases and salts which give water a pleasant taste.

Chemically, water is a simple substance. It is H_2O – a compound containing two hydrogen atoms and one oxygen atom. Yet, it has many amazing characteristics that help it to support and sustain life:

Supersolvent: Water is sometimes described as the 'universal solvent' because it dissolves many substances. The ability of water to dissolve almost everything enables it to carry nutrients through the bodies of plants and animals.

Ability to climb against gravity: Water can move easily through extremely narrow tubes and ooze through invisibly tiny holes. This 'capillary action' lifts water up from under the ground, through the soil to the roots of plants. It then ascends through stems and leaves.

A hearty appetite for heat: Water can absorb more heat than any other substance without considerable rise in temperature. The slow cooling and warming of water prevents extreme climatic changes and protects living things from the shock of abrupt temperature changes.

Ice is lighter than water: Practically every substance contracts as it becomes colder, but water – when cooled below 4°C – expands to have about 10 per cent more volume as a solid than as a liquid. If ice did not float, oceans and bodies of water would be frozen from the bottom up and there would be no living things in them.

The peculiar behaviour of water is the result of the so-called *hydrogen bond* between molecules of water. The hydrogen bond links the positively charged nucleus of a hydrogen atom in one water molecule to the negatively charged electron cloud of a nearby oxygen atom in another water molecule. There is another property of water that makes it unique. It is a polar molecule, which simply means that one part of the molecule has a positive charge and the other negative charge. Other polar molecules can dissolve in water but non-polar molecules cannot. The walls of cells are made of non-polar carbohydrates, ensuring that they will not dissolve in water.

Because of hydrogen bonding considerable energy is needed to force water molecules apart. That is why water has high boiling and freezing points. In ice, hydrogen bonds cause the water molecule to take shape as shown in the accompanying diagram (the dotted lines represent hydrogen bonds). This results in the formation of 'hole', a phenomenon that makes ice lighter than water.

100. Wave

A wave is a periodic disturbance or oscillation travelling in space or in a medium. Wave motion transfers energy from one point to another without the physical transfer of the material. A wave in which vibrations that constitute the wave occur in the direction in which the wave travels is called a longitudinal wave. Sound waves are longitudinal because the particles in the medium vibrate to and fro in the same direction as wave motion. A wave in which vibrations are perpendicular to the direction in which the wave travels is called a transverse wave. Light waves are transverse waves.

To-and-from motion in which a moving particle traces a symmetric path about an equilibrium position is called simple harmonic motion. It happens when the force pulling the particle back to a centre position is proportional to the distance of the particle from the centre. The motion of a simple pendulum (a small object suspended from the end of a light cord) is simple harmonic: it oscillates along the arc of a circle with equal amplitude (maximum displacement) on either side of the equilibrium point.

Simple harmonic motion can be shown as a sine wave (*see diagram*). The highest points on a sine wave are called crests, low points troughs.

The furthest distance an oscillating object moves from the centre of vibrations is called the amplitude of the wave. It is the maximum height of a crest, or depth of a trough. The distance between two successive crests or troughs is known as wavelength. Wavelength is also equals the distance between any two successive identical points on the wave.

Frequency is the number of crests – or complete cycles – that pass a given point in one second. Period is the time required for one complete wave to pass a given point. Wave length is the velocity at which the wave crests appears to move. The equation below shows the relationship between velocity, frequency and wavelength of a wave.

velocity = frequency × wavelength

Phase is a measure of whether a periodic function is 'in step' or 'out of step'. The points on a wave which have the same speed and displacement of movement are said to be in phase; the points which have different speeds and displacements are said to be out of phase.

An imaginary surface in a wave joining all points at which vibrations are in the same phase is called a wavefront. A wave spreading out in all directions from a point source has spherical wavefronts, a parallel beam has plane wavefronts. A line drawn in the direction of motion, perpendicular to the wavefront, is called a ray.

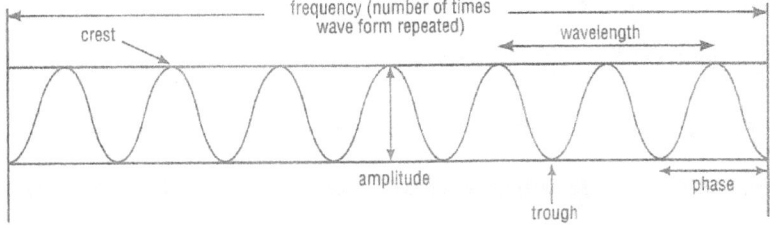

Your understanding of science is incomplete unless you know the difference between science and **pseudoscience** ...

101. Pseudoscience

It is important to know scientific concepts that that are crucial to the understanding of science today. At the same time, it is also important to know ideas that masquerade as science, but have no or little relationship to scientific method. As opposed to science, such ideas are pseudoscience. Theories of real science are continually being added to and updated, but the ideologies of pseudoscience are fixed. Put simply, pseudoscientists believe that their hypothesis can never be wrong, but a real scientist always welcomes new ideas as these ideas give them the opportunity to test their hypothesis in new situations.

The list of popular pseudoscientific ideas and beliefs is rather long. Here's a selection: alien abduction, ancient astronauts, astrology, Bermuda triangle, biorhythms, chiropractic, crop circles, crystal healing, dowsing, extrasensory perception (ESP), graphology, homeopathy, Loch Ness monster, magnetic therapy, near-death experience, numerology, out-of-body experience, palmistry, perpetual motion machine, psychokinesis, pyramid power, quantum healing, telepathy, teleportation, trepanation and UFOs,

In his 1952 book, *In the Name of Science* (which was republished in 1957 as *Fads and Fallacies in the Name of Science*), Martin Gardner, a well-known author of numerous books and a relentless fighter against pseudoscience who died in 2010, launched the modern sceptical movement. In this book he lists five characteristics of pseudoscientists:

- They consider themselves geniuses.
- They regard other scientists as ignorant blockheads.
- They believe themselves unjustly persecuted and discriminated against because recognised scientific societies refuse to let them lecture and peer-reviewed journals ignore their research papers or assign them to

'enemies' to review them.

- Instead of sidestepping the mainstream science, they have strong compulsions to focus on the greatest scientists and best-established theories. For example, according to the laws of science a perpetual motion machine cannot be built. A pseudoscientist builds one.

- They often write in complex jargon, in many cases using terms and phrases they themselves have coined. Even on the subject of the shape of Earth, you may find it difficult to win a debate with a pseudoscientist who argues that Earth is flat.

www.ingramcontent.com/pod-product-compliance
Lightning Source LLC
Chambersburg PA
CBHW071257220526
45468CB00001B/166